EFFECT OF GEOMECHANICS ON MINE DESIGN

PROCEEDINGS OF THE 12TH PLENARY SCIENTIFIC SESSION OF THE INTERNATIONAL
BUREAU OF STRATA MECHANICS / WORLD MINING CONGRESS
LEEDS / 8 - 13 JULY 1991

Effect of Geomechanics on Mine Design

Edited by
A. KIDYBIŃSKI & J. DUBIŃSKI
Central Mining Institute, Katowice

A.A.BALKEMA / ROTTERDAM / BROOKFIELD / 1992

CIP-DATA KONINKLIJKE BIBLIOTHEEK, DEN HAAG

Effect

Effect of geomechanics on mine design / ed. by A.
Kidybinski & J. Dubinski. – Rotterdam [etc.] : Balkema.
– Ill.
Proceedings of the 12th Plenary scientific session of the
International Bureau of Strata Mechanics/World Mining
Congress, Leeds, 8-13 July 1991.
ISBN 90-5410-040-0 bound
Subject heading: mining engineering.

The texts of the various papers in this volume were set individually by typists under the supervision of each of the authors concerned.

Published by
A.A. Balkema, P.O. Box 1675, 3000 BR Rotterdam, Netherlands
A.A. Balkema Publishers, Old Post Road, Brookfield, VT 05036, USA

ISBN 90 5410 040 0

© 1992 A.A. Balkema, Rotterdam
Printed in the Netherlands

Effects of Geomechanics on Mine Design, Kidybiński & Dubiński (eds) © 1992 Balkema, Rotterdam. ISBN 90 5410 040 0

Table of contents

3 *Design considerations and mining technology improvements*

Effects of Geomechanics on Mine Design, Kidybiński & Dubiński (eds) © 1992 Balkema, Rotterdam. ISBN 90 5410 040 0

Introduction

A. Kidybiński
President of the International Bureau of Strata Mechanics, Katowice, Poland

This volume contains 19 papers presented at the 12th Scientific Session of the International Bureau of Strata Mechanics (IBSM) which was held in Leeds, Great Britain, from 8-13 July 1991.

The main theme of this conference was 'Ground Control – Case Studies'. But, as is usual for the Bureau's conferences, most attention was paid to practical recommendations resulting from geomechanical research.

Most papers are coal-mining related, but solutions discussed may in many cases also be applied to metal-ore and salt mining as well as to civil engineering-aimed tunnelling in rock.

The material presented has been divided into three sections, namely:
– on site investigations (7 papers);
– theoretical and laboratory studies (7 papers); and
– design considerations and mining technology improvements (5 papers).

From the point of view of technical aspects considered, the most important items are the following:
– geodynamic phenomena associated with mining (4 papers);
– longwall face technology (3 papers);
– permanent mine workings and tunnels (3 papers);
– modified mining systems (1 paper);
– ground subsidence due to mining (2 papers);
– yield pillars (1 paper); and
– water inflows to mines (1 paper).

We express the hope that this book will be useful both for mine operators and scientists connected with mining engineering.

1 On site investigations

Effects of Geomechanics on Mine Design, Kidybiński & Dubiński (eds) © 1992 Balkema, Rotterdam. ISBN 90 5410 040 0

Controls exerted by dominant parting planes over the deformation of tabular deposits

B.G.D.Smart
Department of Petroleum Engineering, Heriot-Watt University, Edinburgh, UK

ABSTRACT: The paper argues that distinct features termed "dominant parting planes" control the deformation of tabular deposits, particularly with regard to the release of strain energy around an excavation. These features are activiated in shear at some distance from the excavation and may permit subsequent rock failure in a tensile mode due to the associated reduction in lateral confining stresses. Evidence to support the arguement is presented from a number of field observations, including seismicity associated with mine working.

1. INTRODUCTION

The concept of the dominant parting plane was proposed by Smart following a combination of direct observations and measurement of such planes in stratiform sedimentary (primarily coal) deposits.

A dominant parting may be defined as a laterally extensive but thin natural feature parallel to bedding which allows both relative normal motion (separation), and importantly parallel motion (shear) between adjacent strata. It is proposed that both the location and number of such partings together with the relative strengths of the strata that they separate offer a significant control over the manner in which the in situ stresses are concentrated and released around mining excavations, more so than other discontinuities parallel to bedding which simply allow separation.

Thus, for example, overbreak will be minimised in tunnel drivage if a dominant parting is used as the roof horizon, especially if the stratum on the upper side of the parting is relatively strong. Alternatively, overbreak can be excessive if the roof horizon is maintained in close proximity to but below a dominant parting plane, the strata naturally wanting to break up to the parting.

While separation across the parting ultimately causes the "parting" that creates the overbreak, it is likely that even at moderate depth the parting has been activated in shear ahead of the excavation as horizontal stresses in the strata are released. This is particularly true if the excavation is large in relation to the distance between partings e.g., a longwall coalface. This shearing action creates striations in a

layer of debris on the parting, the striations tending to point toward the approaching excavation and the associated release of horizontal stress. As an excavation advances through a sedimentary deposit, the dominant partings therefore control the release of horizontal stress and then having lost all tensile strength subsequently influence vertical movement of strata into the excavation. A photograph of a striaited parting plane surface is shown in Figure 1.

A table of dominant parting planes observed directly is given in Table 1. (After Smart, Olden and Metcalfe). The same paper reports an analysis of both relative vertical and lateral movements measured between roof and floor of a number of longwall coalfaces, as shown in Table 2. Note that the relative lateral movement measured between roof and floor is of the same order as the vertical movement over the supported area of the coalface, and therefore is just as important regarding the redistribution of stresses ahead of and around the coalface. In a more general context therefore, it is argued that strata around a destressed zone will exhibit a tendency to move laterally (i.e., parallel to bedding) toward that destressed zone. For any given stratum, that tendency will be governed by its physical properties and its position relative to the destressed zone. If adjacent strata exist with different tendencies to move laterally, then the potential for shear between them is created. That potential can be released as relative movement parallel to bedding along a dominant parting plane.

Although an analysis of the orientations of the lateral movement reported in Table 2 is not conclusive, it is possible that the direction of movement may be influenced by the orientation of

Table 1 Details of dominant parting planes observed in the vicinity of mine workings. Reproduced from Smart and Crawford 1989

SITE DESCRIPTION	DOMINANT PARTING PLANE LOCATION	OBSERVATIONS REGARDING STRIATIONS, DEBRIS AND MINERALOGY	COMMENTS
1. MALTBY COLLIERY Swallow Wood Seam 24's TG Observations made at ripping lip, above the half-heading.	3.05m above the seam, between a mudstone and a siltstone (upper) stratum.	Striations visible on the underside of the siltstone, pointing into the waste at an angle of 45° to the face. Striation length typically 20mm. 1mm thick debris layer, 70% coverage of shear surfaces.	Parting exerted major control on the deformation of the immediate roof over the half-heading, allowing the immediate roof below it to separate from the upper strata and subside into the heading.
2. MALTBY COLLIERY Swallow Wood Seam 29's TG Junction.	Location similar to above.	25mm thick, highly sheared carbonaceous debris, 100% coverage of shear surfaces.	Site at edge of old faceline, which presumably caused shearing action.
3. KELLINGLEY COLLIERY Beeston Seam 85's MG.	1.42m above seam, between two mudstone strata.	No striations observed.	Parting allows separation of lower strata from upper, governing height of the roadway.
4. BILSTON GLEN COLLIERY Great Seam Roof fall on Coalface, 50m from MG.	3m above coal seam, in mudstone. Sandstone stratum exposed for 3m above the parting plane.	"Scuff" marks containing striations visible on underside of upper stratum. Texture indicated movement towards the waste, at 90° to the faceline. No length measurements possible because of difficult access.	Bridging action of the upper roof strata behind the faceline began above this parting plane. Failure of roof strata forward of the faceline began below it.
5. BILSTON GLEN COLLIERY Great Seam Longwall Coalface.	100mm above the seam, between sandstone strata.	Striations visible in 10mm thick debris, indicating up to 25mm movement towards the waste.	Particle size analysis conducted on this debris.
6. DEEP NAVIGATION COLLIERY Five Feet Gellideg Seam Roof fall on V105 Longwall Coalface.	4.25m above the seam, between mudstone (lower) and siltstone (upper).	Striations visible in 1mm thick debris, indicating up to 25mm movement at 25° to face, normal to steeply dipping joint set in mudstone. Shearing action may be due to "wedging out" of the mudstone.	Coalface abandoned because of roof stability problems encountered when swung almost parallel to joints in mudstones. Mudstone between top of seam and dominant parting plane could not be held in place.
7. DOORNFONTEIN GOLDMINE Mechanised Stope Face.	At hangingwall on top surface of 100mm thick shale band, between two quartzites.	Striations visible in 0.5mm thick highly micaceous debris, pointing towards nearest free surface on stepped face. Striation length 10-15mm. Shearing action may be due to "wedging out" of fractured stope face.	Early "original" energy calculation performed using industrial milling equation estimated 14×10^6 J per m^2 of parting plane dissipated in comminution.

Table 2 Recorded relative lateral movements between the roof and floor on longwall coalfaces. Reproduced from Smart, Olden and Metcalfe

MINE No.	AVERAGE CONVERGENCE (mm)	AVERAGE CYCLE TIME (mins)	AVERAGE MOVEMENT TOWARDS FACE (mm)	AVERAGE MOVEMENT TOWARDS WASTE (mm)	AVERAGE MOVEMENT TOWARDS M/G (mm)	AVERAGE MOVEMENT TOWARDS T/G (mm)
1	16.00	200	1.00	15.90	11.13	0.23
2	10.22	213	1.38	35.86	12.17	4.53
3	9.39	225	2.70	13.73	25.33	5.99
4	-	250	6.72	24.85	13.61	9.87

Fig. 1 Photograph of a dominant parting plane near the roof of a coal seam

maximum horizontal stress as well as direction of advance of the excavation, i.e., the resultant of lateral movement can be made up from a component normal to the direction of face advance and a component parallel to the direction of maximum horizontal stress. This may however be modified locally by the wedging out of blocks created by a jointing system.

Based on the arguments presented above, the controls that dominant parting planes may exert on the deformation of strata around a longwall coalface are illustrated diagrammatically in Figure 2, following an earlier appreciation of their importance reported by Smart and Aziz (1986).

Note that release of horizontal confining stresses in the geological conditions shown permit the propagation of tensile fractures in the roof strata ahead of the coalface. These fractures may subsequently allow downward shear displacement between the adjacent roof blocks if not controlled by the support system. It is postulated that it is the mechanism of shear displacement along dominant parting planes that controls the deformation (subsidence) of the roof strata ahead of the coalface, and as such, these parting planes are responsible for

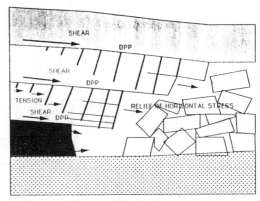

DPP = DOMINANT PARTING PLANE

Fig. 2 The control exerted by dominant parting planes on strata deformation around the longwall coalface

the release and dissipation of strain energy.

The physical properties of sheared parting planes and the effect of that shearing action on the dissipation of strain energy are examined below.

2. THE SHEAR PROPERTIES OF ACTIVATED PARTING PLANES

The physical shear characteristics of discontinuities in various rock types were determined on a servo-controlled direct shear machine shown in Figure 3.

118 mm cylindrical rock specimens were sheared under constant normal displacement control to simulate the action of the "infinite" rock mass in situ. Typical results are presented in Figure 4. This work has been reported extensively previously (Smart and Crawford, 1989), and the results are summarised below.

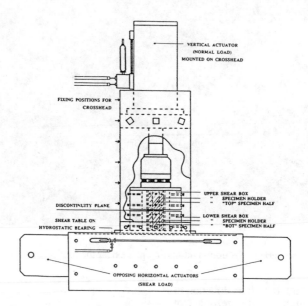

Fig.3 Servo-controlled direct shear machine.Reproduced from Smart and Crawford 1989

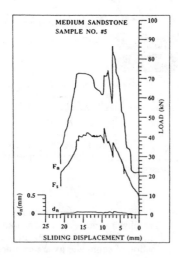

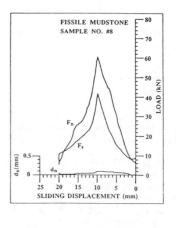

Fig.4 Results from direct shear tests. Fs = shear force, Fn = normal force.Reproduced from Smart and Crawford 1989

The maximum and dynamic coefficients of sliding friction were evaluated for discontinuities in various rock types, and the quantity and size distribution of the shear debris created determined by sieving.

It was found that the size distribution closely followed the Rosin-Rammler law (Sibon, 1980), which can be expressed in the form:

$$\log \log {}^{100}/_R = \log b + \log \log e + n \log x$$

where:

R = % weight oversize retained on a given sieve size,

b, n = constants, so-called "Rosin Numbers",

x = particle size

The constant n is termed the distribution constant as it is a measure of the size range within the debris, and it was found that their was a strong relationship between this constant and the dynamic coefficient of friction as shown in Figure 5. This offers the possibility of estimating the in situ dynamic coefficient of friction for naturally sheared dominant parting planes and other discontinuities from a back analysis of the debris coating their surfaces, as demonstrated in Figure 5.

input into the shearing test was absorbed in comminution, although there was a dependency on rock type in the ratio 1:5:30 for carbonaceous siltstones, medium sandstones and fissile mudstones respectively.

The remainder of the energy in the relatively steady and slow rate shear tests must have been converted into heat. The relative partitioning of the energy input agrees with the estimates made previous (Sibon, 1980). In situ, where the shearing action may be truly dynamic, and where the energy released is derived from the total pre-shear strain energy stored in the strata, energy will also be dissipated in the creation of micro fractures in the vicinity of the shear planes, near vertical macro fractures initiated due to a reduction in the horizontal confining stress, and lastly seismic energy.

Evidence of microfractures are evident in Figure 6, which shows twin thin sections made from the bottom side of the dominant parting plane located at the face of a stope in Doornfontein Goldmine (see Table 1 and Figure10). Note that schistosity has been induced in the direction of shear.

The reduction in strain energy when shear is permitted along dominant parting planes in the vicinity of a gateroad and longwall coalface has been

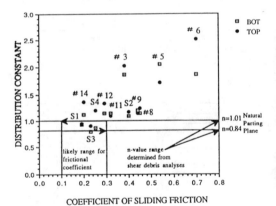

Fig.5 The relationship between the distribution constant n and the coefficient of sliding friction. Reproduced from Smart and Crawford 1989

3. ENERGY DISSIPATION DURING SHEARING OF PARTINGS

An estimate of the energy dissipated in communication during the shearing action was also made, assuming various particle shapes and hence the new surface area created in a given debris. The results showed that less than 1% of the total energy

Fig.6 (a) Thin section of rock from the underside of a dominant parting plane, Doornfontein Mine. Section cut parallel to the striations on the parting plane

Fig.6 (b) Thin section of rock from the underside of a dominant parting plane, Doornfontein Mine. Section cut normal to the striations on the parting plane

estimated using a finite element model as shown in Figure 7. Considering this to be a 1 metre thick section, it can be seen that a significant 4.8 MJ reduction in strain energy occurs when slip is permitted.

4. ADDITIONAL EVIDENCE SUPPORTING THE CONCEPTS OF DOMINANT PARTING PLANES.

4.1 Seismicity

A number of rock failure mechanisms have been suggested (Hasegawa et al., 1990), as the course for seismicity associated with mining, as shown in Figure 8. Failure mechanism (f), i.e., shallow thrust faulting, approximates to the strata deformation mechanisms attributed to the effects of dominant parting planes in Figure 2. There is evidence however of an even closer link between observed seismicity and Figure 2. Teisseyre (1985), suggested that rock deformation process which utilises creep to initiate a rapid release of strain energy would produce the combination of shear and tensile (or implosional) mechanisms thought to occur in both earthquakes and mining-induced seismic events. These mechanisms have been found to

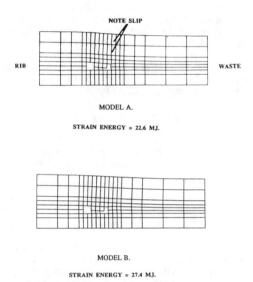

MODEL A.

STRAIN ENERGY = 22.6 MJ.

MODEL B.

STRAIN ENERGY = 27.4 MJ.

Fig.7 The reduction in strain energy predicted by a finite element model of strata around a longwall when slip is permitted on dominant parting planes

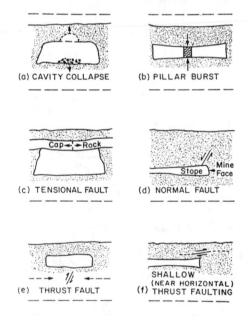

Fig.8 Schematic diagram of six possible ways in which mine-induced seismicity can occur. Solid arrows indicate mine-induced forces on the host rock. Dashed arrows denote ambient tectonic stress. Reproduced from Gibowicz,1990, after Hasegawa et al., 1990.

occur sequentially, or simultaneously but with separation in space.

The combination of shear and tensile mechanisms can be reproduced in the laboratory using the test configuration shown in Figure 9, as proposed by Nemat-Nasser, S. and Horii, H. 1982, and Sileny et al 1986. Note that a tensile fracture is inclined at a steep angle to a shear plane as shown in Figure 2.

A creep process could be generated by shearing along a clay-rich parting plane, while a rougher parting with brittle mineralogy would result in a more marked stick-slip action, both types of process being demonstrated in the results of Figure 4. In either case, the shearing action will release horizontal confining stresses which in turn generates the sudden failure in tension of the strata between dominant parting planes - hence the P waves in the former case and Sand P waves in the latter detected in seismic monitoring.

It is therefore proposed that some of the seismic events associated with mining are initiated by slip along dominant parting planes.

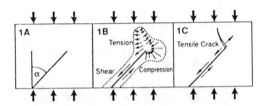

Fig. 9 Test configuration for the reproduction of combined shear and tensile mechanisms. Reproduced from Sileny et al 1986

4.2 Direct measurement of relative lateral movement between strata separated by a parting plane.

While a number of site investigations using borehole instrumention have discovered possible dominant parting planes by accident through shearing of one part of the hole relative to the other, investigations conducted by Legge (1984), were directed specifically at determining the relative lateral movement between strata separated by a parting plane. These were conducted at the stope face in Doornfontein Goldmine, South Africa and typical results are shown in Figure 10, clearly demonstrating the extent to which lateral movement differed in adjacent strata. The thin sections shown in Figure 6 where produced from rock adjacent to what may justifiably be called the dominant parting plane in the hangingwall.

This reasoning also has implications regarding the elimination of large seismic events (rock bursts) when mining in faulted ground. If shear along dominant parting planes is indeed a method of reducing the strain energy in rock ahead of the advancing stope or coalface, then any natural feature that inhibits the shear will results in a storing of strain energy - to be released subsequently in a large event. One such natural feature is a fault, which will displace dominant parting planes and therefore lock in strain energy when approached perpendicular to strike. When mining along strike however, the dominant partings planes are continuous in the direction of strain relief, i.e., back into the waste, and therefore energy dissipation by shear through continual shearing of the partings should be uninterrupted, reducing the likelihood of large seismic events.

5 . CONCLUSIONS

The following points are proposed:-

(i) Shear along dominant parting planes reduces the strain energy in a stratiform deposit as it reacts to the creation of a stress-relieving zone within it.

(ii)   The shearing action produces a layer of rock debris with striations on the parting plane. The direction and extent of shear movement can be determined from the striations.

(iii)   The resultant of shear movement definitely has a component directed toward the stress relieving zone, it may also have a component in the direction of maximum horizontal stress.

Alternatively, the shear movement direction may be modified by the wedging out of blocks formed by a jointing system.

(iv)   The size distribution of the shear debris obeys the Rosin-Rammler law. The distribution constant of that law is related to the dynamic coefficient of friction.

(v)   Microfractures may also be induced in the rock immediately adjacent to the shear debris. The orientation of the microfractures are related to the direction of shear.

(vi)   The combination of stick-slip and fracturing intension mechanisms produced, it is proposed, by slip along dominant parting planes may account for some of the seismic events recorded from around mine workings in tabular deposits.

Accepting the importance of the dominant parting planes in controlling strata deformation, it is proposed that efforts are made to detect their existence and quantify their effects in site investigations. Their impact on the behaviour of the

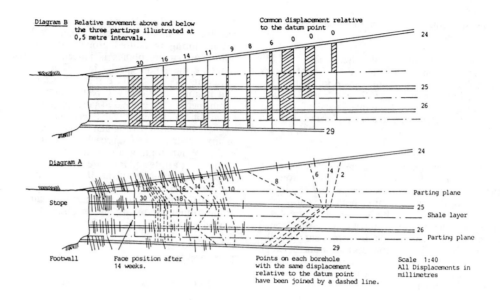

Fig. 10 Results from extensometers showing relative lateral movement between adjacent strata at a gold mine stope face. After Legge, 1984.

rock mass makes them much more important than other discontinuities or fractures parallel to bedding which are often used simply to evaluate RQD. Logging of core data should therefore include a specific attempt at identifying them.

REFERENCES

Hasegawa H.S., Wetmiller R.J., Gendzwill D.J.,1989. Induced Seismicity in Mines in Canada - AN Overview. In Seismicity in MInes, ed Gibowicz S.J., Pure Appl. Geophys. 129 423-453.

Legge, N. B., 1984. Rock Deformation in the Vicinity of Deep Gold Mine Longwall Stopes and its relation to Fracture. PhD Thesis, University College Cardiff.

Nemat-Nasser, S., and Horii, H., 1982. Compression-induced nonplanar crack extension with application to splitting, exfoliation, and rockburst. J. Geophys. Res. 87, 6805-6821

Smart B.G.D., Aziz N.I. The Influence of Caving in the Hirst and Bulli Seams on Support Rating. 1986. Proc Symp Ground Movement and Control Related to Coal Mining. Aus IMM Illawarra Branch pp 182 -193

Smart B.G.D., Olden P.W.H., Metcalfe K. Consideration of the Lateral Faces Generated by Powered Supports. To be published in the Mining Engineer.

Smart B.G.D., Crawford B.R.A., 1989. The Nature and Properties of Dominant Parting Planes Activiated in the Extraction of Tabular Deposits" SERC Report - No GR/D 3731.

Sibon R.H., 1980 Power dissipation and stress levels on faults in the upper crust. JGR 85 (B11) pp 6239-6247

Sileny J., Ritsema, A.R., Csikos I., Kozak J.,1986. Do some shallow earthquakes have a tensile source componant ? In Physics of Fracturing andSeismic Energy Release, eds Kozak J., and Waniesk L. Pure and Appl. Geophys. 124 825-840.

Teisseyre, R., 1985. Creep Flow and Earthquake Rebound: System of Internal Stress Evolution. Acta Geophys. Pol. 33 pp 11-23

Effects of Geomechanics on Mine Design, Kidybiński & Dubiński (eds) © 1992 Balkema, Rotterdam. ISBN 90 5410 040 0

Development of mining induced seismicity in relation to the advance of coalfaces in Karviná, part of Ostrava-Karviná Coal Basin

P. Konečný
Mining Institute of Czechoslovak Academy of Sciences, Ostrava, Czechoslovakia

ABSTRACT: Evaluation of induced seismicity development - of number and energy of seismic phenomena - is applied in Ostrava-Karviná Coalfield already for some time as one of the methods of assessment of rockburst occurrence risk. Until recently there was no method available which should enable to adjoin information on mining intensity to regional seismicity development evaluation in an adequate manner. The proposed procedure of evaluation of mining intensity makes possible both the definition of area in which a correlation between worked-out and induced seismicity is obvious and a quantitative determination of this correlation. This determination possesses also a prognostic importance. The methodic procedures are completed by examples of seismicity development analyses prior to rockburst occurrence.

1 INTRODUCTION

The mining-induced seismic events are common occurrence in the mines of the Ostrava-Karvina Coal Basin. Figure 1 shows the quarterly coal production in 1989-1990 (total and from seams with rockbursts exposure) versus recorded seismic events with magnitudo greater then 0.4. Frequency of these events related to their magnitudo is plotted on Figure 2 and 3. Some of these events - rockbursts - have serious consequences for the operation of the mines. Therefore, emphasis is put on the improvement of the level of interpretation of geophysical data and on the processing of the great quantities of information provided by the seismic networks.

2 PRIMARY PRESUMPTIONS

Evaluation of induced seismicity development in connection with mining activity within regional concept, aimed at prognostic results in relation to rockburst control problems is based on the following presumptions (Rudajev 1986, Konečný 1990):

1. Induced seismicity phenomena reflect the ground disturbance process. The set of so called bulletin data on seismic phenomena - time of occurrence, coordinates of coal area, intensity of rockburst phenomena - describes at the same time the deve-

lopment of ground disturbance process and the development of its parts in time and space.

2. Ground disturbance is a resultant of mutual effect between the stress field and strength in every point of ground where after exceeding a limit strength value a disturbance occurs. The main factors in

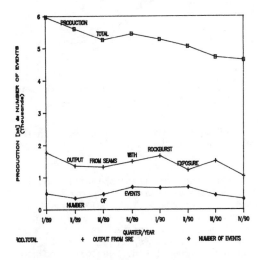

Fig.1 Mining induced seismicity and coal production in Ostrava-Karviná Coal Basin

this process are obviously primary and secondary stress fields (they determine
a resulting stress field) and also geologic and geomechanical structures of the
ground (especially rock strength is determined).

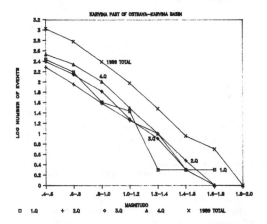

Fig.2 Frequency of seismic events in the
year 1989

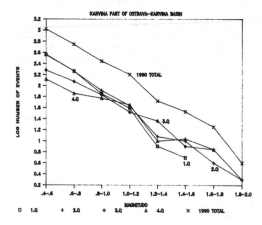

Fig.3 Frequency of seismic events in the
year 1990

3. In regard to time variability of the
main factors affecting the ground disturbance the secondary (induced) stress field
is characterized by the greatest dynamics
as it originates and varies due to mine
working advancements. The other factors
- primary stress field, geologic and geomechanical structures - remain relatively
constant during all simultaneous periods

of face life, level life and mine life. If
a change due to human intervention occurs
it is then usually a single unrepeated
change with long-term constant effects.
4. In compliance with the above - mentioned is also a sensibility of the factors
mentioned to human activity effects. As
the advancement of mine workings is a result of anthropogene activity and is therefore determined by human factor primary
stress fields, geologic and geomechanical
structures are natural factors with very
limited sensibility range.
It can be concluded from the above
- mentioned consideration that for a knowledge of ground disturbance development
laws which apply during mining activity in
the given area it is necessary to analyse
the correlation of two variables:
- of independently variable mine working
advance as induced stress source,
- of depending variability of induced
seismicity which reflects the ground disturbance development on the background of
relatively time independent, constant geologic and geomechanical structures of rock
ground area concerned.

3 RECOMMENDED METHOD OF VARIABLE VALUE
CALCULATION - CALCULATION OF SEISMICITY
DEVELOPMENT AND MINE WORKING ADVANCE (WOR-
KING INTENSITY)

To enable required analyses of correlation
between mining activity advance and induced seismicity the information on both variables must be processed in a comparable
manner. As a base the reporting on seismic
phenomena has been considered - their number and intensity in 250 x 250 m squares.
An agreed network was lain on the area of
Ostrava - Karviná - Coalfield which comp-
lies with the locating capacity of seismic
networks of the Coalfield as well as characteristic dimensions of longwall faces
of OKR (i.e. Ostrava - Karviná - Coalfield). It should be added that an increase
of network density could be justified only
by local problem solution (but always with
taking regard to the accuracy of applied
data). In a regional scope however the
number of data processed increases excee-
dingly.
The development of ground disturbance
process in the area concerned has been
described by mentioning number and intensity of seismic phenomena occurring in
a selected time interval in individual
250x250 m squares. The phenomena could be,
of course classified into categories of
intensity or eventually according to focal
distance or other additional criteria.

For an adequate pursuance of mining advancement the author of this paper has developed a new working intensity evaluation method. (Konečný 1990). The new method of mining intensity evaluation correlated to the induced seismicity development is based on the presumption that a change of stress-strain condition in rock strata affected by mining activity causing ground disturbance and thus also and consequently occurrence of seismic phenomena would be the more intensive with increasing worked-out areas in the given time span. At the same time it is necessary to take into consideration not only the proper worked-out area but also the whole gob area.

Now it should be considered that in 1988 according to OKR Annual Review the average daily drivage advance amounted to 4.46 m per day at a cutting cross section of 14.1 m², representing a daily output of 63 m³ of cut rock. On the other hand in the same year 1988 the average longwall face advance with application of powered support was 1.85 m per day or 217 m² worked-out area per day which means about 520 m³ of output per face and day at a seam thickness of about 2.4 m. In this situation it can be proved (Konečný 1990) that daily output volumes in areas without stress amount to about 130 m³ in roadway development while in longwall faces daily outputs reach up to 3640 m³ per day and face. A twenty eight times output difference has resulted both from the nature of face entries which are maintained and from the caving of worked-out area of longwall face (caving gob). When working with stowing the output is approximately halved.

From the above-mentioned data results an important conclusion that the main source of induced stresses is longwall face advancement while the mining intensity evaluation can be based on assessment of worked-out volume during winning in a selected time interval.

For an evaluation it is necessary to take into consideration the location of face in rock strata in which the winning is realized. In connection to methodology of seismicity development processing a 250 x 250 m network is applied. The mining intensity

$$I_{jk} = p_{jk} \cdot m_{jk} \cdot k_z \quad (1)$$

where
I_{jk}....mining (working) intensity in the square j,k (cm),
p_{jk}....percent proportion of worked-out area (%) in j,k square,

m_{jk}....mean worked seam thickness in square j,k (m),
k_z.....coefficient of gob filling
$k_z = 1$, in case of longwall winning with caving of gob,
$k_z = 0.5$ in case of longwall winning with stowing of gob.

The calculated mining intensity I_{jk} numerically at the same time indicates what seam thickness (cm) should be worked in the square pursued if the whole area of the square would be worked out.

By means of I_{jk} indices the mining intensity in a given time interval in the whole area of region concerned can be expressed.

The I_{jk} index is assessed by reading areas worked-out in a given time interval from mine (seam) maps (Figure 4).

In order to express some other influences and to evaluate the mining intensity in regard to mining depth and worked seam also other facts are included apart from detection of worked-out areas as indicates Table 1 in which a set of input data for mining intensity evaluation and an example of data actually read are mentioned.

For processing the results programs for PC XT/AT computers have been elaborated. The information can be printed out with an arbitrary classification structure (according to time interval, mining depth, worked seam, caving or stowing of gob).

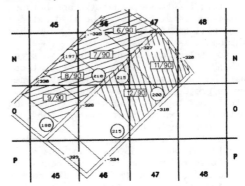

Fig 4 Scheme for reading worked-out areas from mine maps

4 REGIONAL EVALUATION OF INDUCED SEISMICITY IN CORRELATION TO MINING ACTIVITY

It is necessary to perform the evaluation of induced seismicity development in correlation to mining in a regional scope (proportionally adapted also to local problem control) in the following two steps:

13

Table 1 - Data set for mining intensity evaluation

Colliery	year	month	square No	letter	seam (name)	seam thick [m]	gob filling $k_z=1$ or $k_z=0.5$	depth [m]	worked out area [%]	remark
XA1	1990	7	45	N	37c1	2.1	1	-325	3	
	"	7	46	M	"	2.1	1	-325	3	
	"	7	46	N.	"	2.1	1	-325	53	
	"	8	45	N	"	2.0	1	-328	31	
	"	8	45	O	"	2.0	1	-328	5	
	"	8	46	N	"	2.0	1	-328	17	
	"	8	46	O	"	2.0	1	-328	7	
	"	9	44	O	"	1.9	1	-330	2	
	"	9	45	N	"	1.9	1	-330	8	
	"	9	45	O	"	1.9	1	-330	43	
	"	11	47	N	"	2.0	1	-322	63	
	"	11	47	O	"	2.0	1	-322	7	
	"	11	48	N	"	2.0	1	-322	10	
	"	12	46	N	"	2.1	1	-320	16	
	"	12	46	O	"	2.1	1	-320	25	
	"	12	47	N	"	2.1	1	-320	12	
	"	12	47	O	"	2.1	1	-320	36	

First step includes definition of the area in which a correlation between working and induced seismicity (and thus also ground disturbance) can be observed. Within this first step the linkage between mining and seismicity can be also defined as it will be further discussed. For this purpose it is necessary:

- to select an adequate time interval within which the individual evaluations shall be performed. A month interval appears as proper for regression analyses, for actual development processing it will be rational to reduce this interval for instance to a single week,

- to elaborate maps of seismic irradiation in the area concerned and its environment including numbers of phenomena occurred and total energy for a selected time interval and 250 x 250 m squares. Depending on the nature of problem solved it is possible to process such information also by means of various filters - most frequently with classification according to intensity of phenomena occurred,

- to elaborate maps of mining intensity in the same areas and same time intervals as above. Also here an adequate filter according to the nature of problem can be applied. The applied filter may classify according to seam identity (or seam group), or to worked seam height, deposit depth, gob handling (stowing, controlled caving) or combination of gob handling methods,

- by comparison of both groups of maps to define the area in which a causal connection between mining and seismic phenomena occurrence is proved. Such maps and defined areas should be compared with geological and geomechanical maps (background) with the aim of evaluation the influence of natural conditions (geological and geomechanical) on formation of the defined area.

In the second step a quantified comparison between mining intensity and induced seismicity in a defined area is performed where the disturbance development is characterized by seismic phenomena frequency and their energy content.

The application of adequate filters is purposeful also here, for instance for demarcation of affected areas. In this second step correlations can be revealed which are applicable for evaluation of eventual rock burst risk in the area or eventually of risks of other anomalous effects of rock pressure.

5 EVALUATION EXAMPLES

An exemplary evaluation of development of mining advance and induced seismicity with regard to geological and geomechanical conditions based on the above - mentioned method was performed for localities in which in 1988 rockbursts or mictrotremors occurred. It was on the one hand a part of Doubrava mine (microtremor on March 3rd and rockburst on October 18th) and on the other hand a zone at the boundary of ČSA and Darkov Collieries (microtremor on June 29th in Darkov Colliery, microtremor on November 6th in ČSA Colliery and rockburst on December 19th in Darkov Colliery.

5.1 Definition and characteristics of areas-zones based on comparison of mining intensity and induced seismicity

For definition of areas-zones according to the above mentioned method comparisons have been elaborated expressing the correlation between mining intensity and the number of seismic phenomena occurred in 1989 (see Figure 5, Figure 6). It is obvious that rock bursts in Doubrava Colliery belong to a single, separate area. The rock bursts studied in ČSA and Darkov Collieries, i.e. in two underground mines, however appear in view of seismicity development and thus also ground disturbance to be linked to the same single area demarcated by Jindřich fault in the North, by Dora fault from the South, by Barbora fault from West and by Gabriela fault from the East (see Fig. 6).

5.2 Quantification of induced seismicity in correlation to to mining intensity

The mining intensity, number of seismic phenomena occurred as well as irradiated

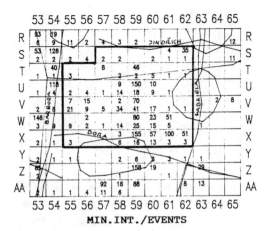

MIN.INT./EVENTS

Fig 6   Mine ČSA and Darkov   - Intensity of mining and number of events (1988)

seismic energy in the area pursued (see Figure 7 resp. Figure 8) were evaluated and graphically plotted in monthly intervals. The months in which rockbursts or microtremors occurred are marked by stars (*).

It is evident, that:
- by adding mining intensity data to the information on development of number of phenomena and seismic energy liberation it is possible to obtain an idea of the ground reaction to mining activities,
- a situation can be considered as normal in which the development of mining intensity corresponds to the development of seismic activity (with increase of mining the energy liberated at ground disturbance increases and with a decrease of mining this energy also decreases),

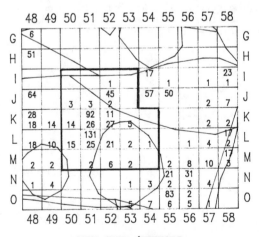

MIN.INT./EVENTS

Fig. 5 Mine Doubrava - intensity of mining and number of events (1988)

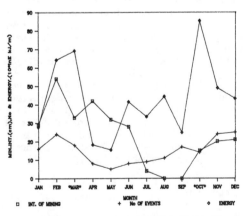

Fig.7 Evaluation of induced seismicity in Doubrava Colliery

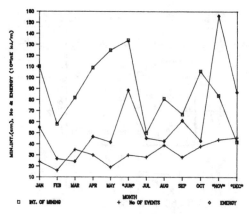

Fig.8 Evaluation of induced seismicity of the area of ČSA and Darkov Collieries

- an anomalous development - in which in spite of mining intensity increase the seismic energy liberated in the given area decreased - resulted in ČSA - Darkov Colliery region into a subsequent intense energy liberation in form of rockburst phenomena (see Figure 8),
- a situation must be considered as anomalous in which in spite of mining intensity decrease the magnitude of irradiated seismic energy increases. At Doubrava Colliery such situation climaxed by a rock burst phenomena in March 1989 (see Figure 7).

Prior to the rock burst of October 1989 in the same colliery the situation was characterized by absence of mining activity, however the decrease of mining and increase of irradiated seismic energy represented again an anomalous situation between July and August and the following start of mining in October caused a rock burst occurrence. It seems that much was indicated by a frequent occurrence of seismic phenomena in the given area in spite of mining pause in August and September. Potential stress and unbalance conditions were thus indicated,
- unbalanced stress conditions in both cases (in Doubrava Colliery especially in the second half of year) are also testified by the fact that in spite of generally descending trend of mining intensity the frequency and number of seismic phenomena is globally increasing.

6 CONCLUSION

The evaluation procedure of induced seismicity in correlation to mining activity in regional scope applied for prognostic purposes consists of the following two steps:
- definition of area in which a causal correlation between mining advance and seismicity phenomena occurrence is evident,
- quantification of correlations between mining intensity and seismic activity with application of various filters emphasizing geological and geomechanical or eventually other technology influence factors.

From the correlations obtained the development of ground disturbance process can be appreciated or eventually it further development can be predicted.

The method presented is designed for obtaining and evaluation of information within a regional scope. When applied for local problem solutions it must confront the knowledge of development of mining induced seismicity with results of geomechanical, geodetic, geophysical and other adequate methods (Rakowski, Konečný 1986).

REFERENCES

Konečný, P. 1990. Disturbance of rock strata and induced seismicity when working seams in the Karviná part of OKR in relation to rockbursts, Habilitation thesis Mining Institute of Academy of Sciences - Mining University in Ostrava
Konečný, P. et al. 1990. Liberation of seismic energy due to mining activity. Final Report DU II-6-6/05 Mining Institute of Academy of Sciences in Ostrava
Rakowski, Z., Konečný, P. 1986. Physical model of rockburst zone. Partial Report, Mining Institute of Academy of Sciences in Ostrava
Rudajev, V. 1986. Rockburst seismics. Doctor thesis UGG Institute of Academy of Sciences in Prague.

Effects of Geomechanics on Mine Design, Kidybiński & Dubiński (eds) © 1992 Balkema, Rotterdam. ISBN 90 5410 040 0

Water inflows and weightings above longwall workings

G.C. Xiao & I.W. Farmer
Ian Farmer Associates, Newcastle upon Tyne, UK

ABSTRACT: A micro-computer database, incorporating historical data from British Coal mines, to study the factors affecting inflows from proximate aquifers into longwall workings, showed that the most important factor affecting mine water inflows was the cover strata lithology, particularly within the 30m of roof strata. When this comprised principally stronger sandstones, there was a tendency for periodic inrushes associated with face weighting. This phenomenon is explained by the collection of water in bed separation cavities, where free water connected to a high pressure head aquifer source may be trapped. This finding attributes the difficulty in predicting mine water inflows to the changeable nature of cover strata lithology.

1 INTRODUCTION

In the past, the principal factor governing water inflow into mine workings was considered to be the depth of cover beneath an aquifer or water source. Garritty (1981, 1982, 1983) in a study of water inflows into longwall workings beneath the North Sea and the Permian aquifer in the Northumberland and Durham coalfields of North East England, showed that there was a high risk of inflows when:

1. Cover to the sea bed or the base of Permian was less than 140m or 105m respectively (Figure 1).
2. Faults with throws greater than 1m connected the workings to the water source.
3. The competence of the immediate roof strata was relatively high – particularly where the first 20-45m contained less than about 35% of mudstones (Figure 2).
4. Inadequate roof control allowed major roof breaks to form – usually associated with a major roof weight.

Garritty's work is the most thorough investigation of water inflows in British Coal Mines and although he did not seek to explain the inflows in terms of specific mechanisms, these can be inferred from his conclusions and the coal face geometry. Thus inflows occur during strata fracturing following caving at a time when major roof breaks occur in strong rocks often accompanied by weighting. This induces continuous fractures in the cover which also connect to the water source. This is exacerbated by faulting and ameliorated by the presence of weaker and less permeable rocks which fracture easily and recompact quickly.

It is, unfortunately, not an easily modelled prototype situation and when similar inrushes occurred during the initial stages of development in the Selby Coalfield (Wilby, 1990), and similar strata were postulated for the proposed new Asfordby mine (Hindmarsh, 1990), it was decided to develop Garritty's work in a more systematic way using an aquifer database compilation (see Xiao et al, 1990).

2 INFLOW DATA

Initially data from two mines in the

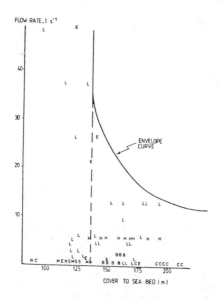

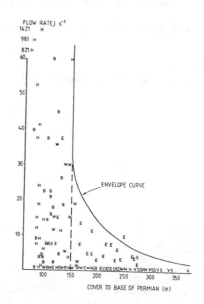

Fig.1 Inflow into workings (a) beneath the seabed in the Northumberland coalfield, (b) beneath the Permian aquifer in the Durham coalfield (after Garritty, 1981)

Note that in this figure and Fig.2, data points are designated by the initial letters representing the mine. In Northumberland, the main mines are Ellington (E), Lynemouth (L), Bates (B) and Broomhill (H). In Durham, the main mines are Blackhall (B), Easington (E), Horden (H) and Westoe (W)

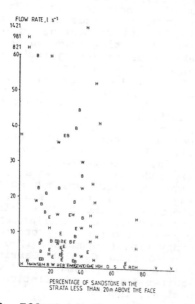

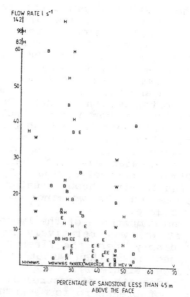

Fig.2 Effect of the percentage of sandstone strata in the Coal Measures sequence (a) 20 and (b) 45m above longwall faces in the Durham coalfield on flow from the Permian aquifer (after Garritty, 1981)

North East coalfield, Horden and Blackhall, were examined. These are two of the best documented mine inflow cases in recent years and are included in Figures 1 and 2 as "H" and "B".

Water inflow at Horden mine mainly occurred in the High Main seam. This was worked under the North Sea with less than 100m cover to the base of the Permian aquifer. Boring from the seam encountered a static head of 228m at the base of the Permian.

In the early 1960s several longwall faces were worked encountering only minor water feeders. However, after 1963 roof water caused delays on two adjacent 210m longwall faces at cover depths below the base of the Permian of 75 and 90m. Subsequent abandonment of a longwall face with flows of 41 l/s led to a decision to develop 65m retreat faces with 50m pillars, but while some of these encountered only minor feeders, others encountered heavy feeders up to 50 l/s. Some of these were associated with faulting; some with coarse saturated Coal Measures sandstones.

At adjacent Blackhall mine, the Low Main seam was worked at about 100m below the base of the Permian aquifer. Inflow events were usually associated with severe face weighting and after several 200m longwall faces had experienced heavy inflows, narrow faces were worked with substantial pillars. Some of these faces also encountered water problems.

Figure 3 summarizes inflow data at Horden mine, which shows (a) the presence of the 'D' sandstone which covering the whole area and incropping to the base of Permian; and (b) that the percentage of incompetent rocks of the roof strata are rather low over the area, particularly within 30m of the roof strata.

Figure 4 summarizes inflow data at Blackhall mine, which shows, except for those specifically associated with faults, a good correlation between inflows and the low percentage of incompetent rocks, particularly within the 30m of roof strata above the seam. The majority of inflows occurred to the north of Blackhall mine (to the right of the graph) where the cover distance to the base of Permian are on the higher side and the percentage of incompetent rocks on the lower side.

The water inflows in the A block at Wistow mine are the most notable such events of recent years in Britain. The first two longwall faces at a cover depth below the Permian aquifer base of 77m had respective inflows of 121 and 11 l/s. The inflows followed heavy weightings and subsequent faces were changed to shortwall faces, with a panel width of 46m. When one of these experienced an inflow of 136 l/s the length was successfully reduced to 37m, but the block was eventually abandoned as uneconomical. Subsequent workings in C block at a depth of 200m below the base of the Permian were water free.

Figure 5 summarizes inflow data at Wistow mine. Figure 5a shows the low cover (around 80m) to the base of Permian over the 'A' block and the high cover (more than 200m) over the 'C' block and it also shows the narrowed panel width and reduced extraction height of later workings towards the west of the 'A' block after the water flushes occurred on the two wide panels (A1 and A2). Figure 5b shows that the cover lithology of the immediate roof strata fluctuates from one borehole to another, particularly from the boreholes in the 'A' block. The fluctuation is clearly in favor of the water occurrences.

The ameliorating effect of weaker and more impermeable rocks can be seen by looking at data from Clifton and Cotgrave mines in Figure 6, which plots the same data as Figures 3b, 4 and 5b and which shows the high percentage of incompetent rocks in the roof strata over the workings at very shallow depths beneath the Permian aquifer.

Situated on the western side of a relatively undisturbed Carboniferous sedimentary basin and dominated by mudstone sequences within the main coal bearing succession, the South Nottinghamshire coalfield has rarely been affected by water inflows – even when working at very shallow depths beneath the Permian aquifer.

At Cotgrave mine several longwall faces in the Deep Hard seam have

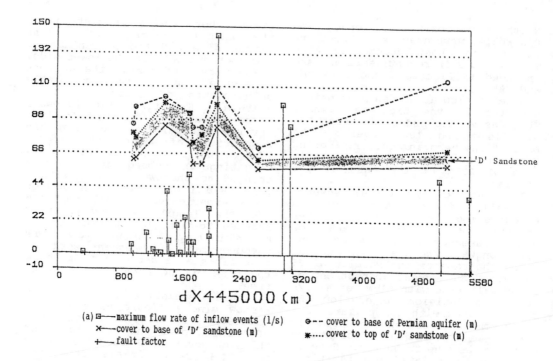

(a) ⊟——maximum flow rate of inflow events (1/s)   ⊙-- cover to base of Permian aquifer (m)
   ✕——cover to base of 'D' sandstone (m)   ✱····cover to top of 'D' sandstone (m)
   +——fault factor

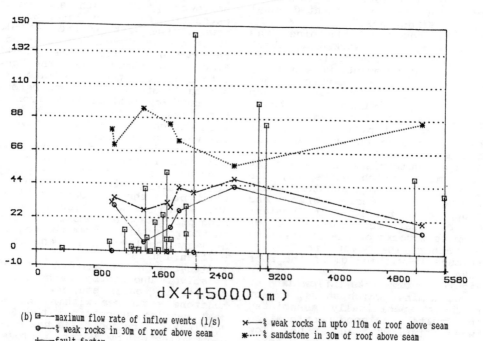

(b) ⊟——maximum flow rate of inflow events (1/s)   ✕——% weak rocks in upto 110m of roof above seam
   ⊙——% weak rocks in 30m of roof above seam   ✱····% sandstone in 30m of roof above seam
   +——fault factor

Fig.3 Historical inflow data at Horden mine. Referenced to the National Grid Line of X=445000 in a west-east direction.

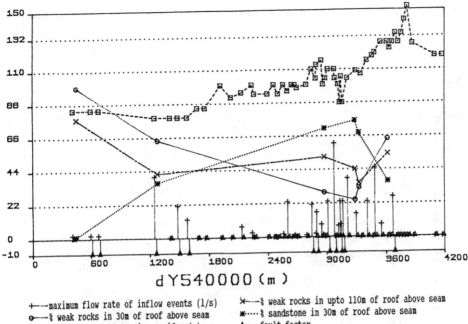

+——maximum flow rate of inflow events (1/s)   ✕——% weak rocks in upto 110m of roof above seam
⊙——% weak rocks in 30m of roof above seam   ✳·····% sandstone in 30m of roof above seam
⊟——cover to base of Permian aquifer (m)   ▲——fault factor

Fig.4 Historical inflow data at Blackhall mine. Referenced to the National
Grid Line of Y=540000 in a south-north direction.

been worked 45m below the Bunter Sandstone, including one district which stopped at a 5m fault hading away from the face. After a few months this became damp and minor water flushes were experienced in methane boreholes. Similar experiences were obtained at nearby Clifton mine. In both cases mudstones comprised 60-70% of the immediate roof and of the full sequence to the base of the Permian.

Asfordby mine is to be developed to extract the Deep Main seam under the Permo-Triassic water bearing strata. Geological data from Asfordby mine, obtained from the Geology Branch of the HQTD, British Coal, in the form of borehole logs, had been input into the Aquifer Database for an assessment of its risk of water inflow. Examination of the data indicated that the aquifer situation at the Asfordby mine is similar to that at the Wistow and Cotgrave mines. And the latter two mines have the contrasting situations as discussed above.

Figure 7 shows that the cover distances over the assumed workings

of the Asfordby mine are no better than those over the workings of the Wistow and Cotgrave mines. It also shows that the cover lithology of the roof strata over the assumed workings of the Asfordby mine is similar to that over the 'wet' workings in the 'A' block of the Wistow mine. The conclusion from the graphical analysis of the geological data from Wistow, Cotgrave and Asfordby and from analogy with mine water problems at Wistow and Cotgrave mines is that the workings at Asfordby mine would risk water inrushes.

3 BED SEPARATION

Hoare and Whitworth (1978) and Elliott (1978) amongst others have proposed a relatively simple explanation for water inrushes in the Cannock Chase coalfield. Following an inrush at West Cannock No.5 mine, boreholes were drilled upwards from the seam and water flows were observed at heights of 18, 38, and 44m in one borehole and 38 and 69m in a second borehole. In

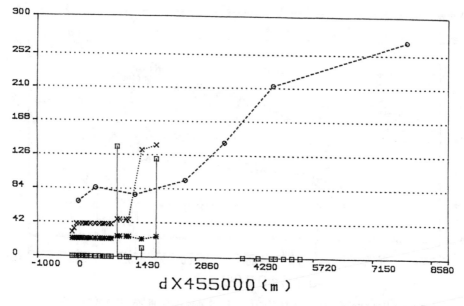

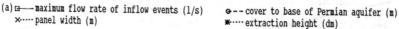

(a) □——maximum flow rate of inflow events (l/s)   ⊙-- cover to base of Permian aquifer (m)
    ×····· panel width (m)   ✳····· extraction height (dm)

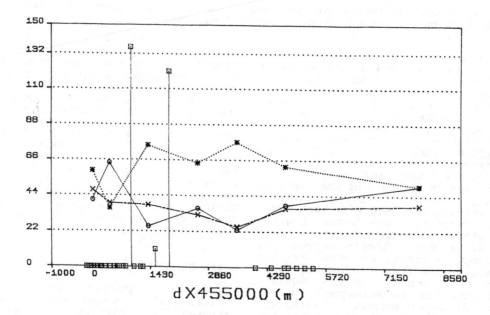

(b) □——maximum flow rate of inflow events (l/s)   ×—— % weak rocks in upto 110m of roof above seam
    ⊙—% weak rocks in 30m of roof above seam   ✳····% sandstone in 30m of roof above seam

Fig.5   Historical inflow data at Wistow mine.   Referenced to the National
Grid Line of Y=455000 in a west-east direction.

22

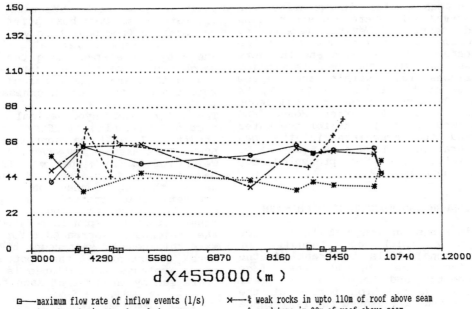

Fig.6 Historical inflow data at Clifton and Cotgrave mines. Referenced to the National Grid Line of Y=455000 in a west-east direction.

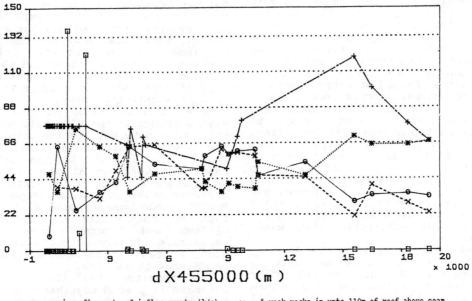

Fig.7 Historical inflow data at Wistow, Clifton and Cotgrave mines and geological data from the proposed Asfordby mine for comparison. Referenced to the National Grid Line of Y=455000 in a west-east direction.

each case water in quantity was observed only where bed separations had occurred. Several de-watering borehole programmes were subsequently initiated and in every case but two the bed separation horizons fall within the narrow zones at the levels of 21, 37, 54 and 72m above the roof of coal seams (see Figure 8, after Hoare and Whitworth, 1978). As for how water comes to accumulate in the bed separations, Figure 9 (after R. Elliott, 1978) is presented for consideration.

4 INRUSH AND WEIGHTING MECHANISM

Weightings on longwall faces occur when a thick bed (usually of sandstone) fails to break at the caving line behind the face supports, and hangs in the goaf until the reaction forces at the sides and face exceed the resistance of the bed. The resultant rapid energy release and accompanying face closure, unless designed into the yielding characteristics of the face supports, can cause damage and lost production. The most severe weightings are usually at the beginning of a face, when a double span collapse may be expected, or where water at pressure may be present in the proximate strata. In the latter case there may be the additional hazard of water inrushes.

Under normal circumstances, when a face has been established, and provided it is long enough to simulate two-dimensional conditions normal to the face line, fractures induced by the front abutment stress, will be sufficient to establish a satisfactory caving regime. Abnormal circumstances may be created by narrow faces, thick nether roof beds and, particularly, thick beds associated with water pressure.

Consider for instance a typical Coal Measures sandstone with a laboratory strength of 85 MN/m² measured on a 25mm specimen. Then for say a 10m thick bed, this will reduce in inverse proportion to the square root of the dimension (Farmer, 1984), giving an estimated rock mass strength of about 5MN/m². Although Boundary or Finite Element

analysis is more accurate, an estimate of maximum beam stress can be obtained from beam theory:

$$\sigma_x = qL^2/2t^2$$

Where L is the span and q the body weight equal to τt per unit length and width where τ is the rock unit weight and t the beam thickness.

Thus for the hypothetical beam section, assuming $\tau = 25kN/m^2$ (156 lbs/ft²), L will be approximately 60m. This assumes, however, a body weight of 0.25MN/m² per m. If the beam is subject to a head of 250m of water through connection to an aquifer, this increases q by an order of magnitude and, in this case, either the span is reduced, or the thickness increased, for the same thickness or span respectively. More importantly, the potential energy released at collapse is also increased by an order of magnitude.

5 CONCLUSIONS

The analytical results from the Aquifer Database (Figures 3 to 6) show consistently a correlation between heavy water inflows and low percentages of incompetent (impermeable) rocks within the roof strata, particularly within the 30m of immediate roof strata. The results also indicate that the cover distance to aquifer is important only if there are impermeable rocks present within the cover.

This is illustrated at the Low Main workings at the Blackhall mine, which were very wet despite cover to the base of Permian of more than 100m, (Figure 4). On the other hand, the workings at the Clifton and Cotgrave mines were virtually dry with cover as low as 45m to the Bunter sandstones (Figure 6). The fact of the matter is that at Clifton and Cotgrave mines the percentage of incompetent rock was high, whereas at Horden, Blackhall and Wistow it was low.

The results from the Aquifer Database if pieced together with the bed separation theory (Figure 8), the pattern of strain zones and water migration above a working panel (Figure 9) and the weighting mechanism (Section 4), they form a clear picture of the basic

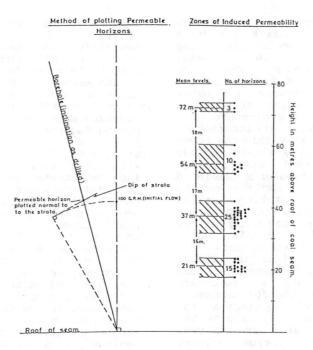

Fig.8   Permeable Horizons encountered in de-watering boreholes in the Cannock Chase coalfield (after Hoare and Whitworth, 1978)

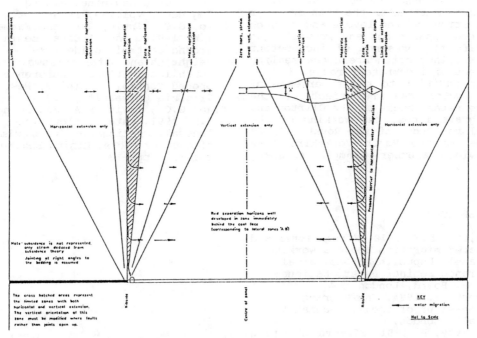

Fig.9   Strain zones and water migration paths into bed separation cavities above a working panel (after Elliott, 1978)

25

mechanisms of mine water inflow into a working panel: Figure 8 provides a path for water movement through the limited zones with both horizontal and vertical extension (the cross hatched areas); the bed separation theory offers reservoirs for water accumulations; the Aquifer Database shows that it is the presence of a certain percentages of incompetent (impermeable) rocks within the cover strata, particularly within the 30m of immediate roof strata that determines whether water accumulated in bed separation cavities can be sealed off from entering working panels; and finally the weighting mechanism explains how water inflow occurs and why it is often associated with strata weights.

The cover to aquifer is further complicated by the presence of Coal Measures sandstones as secondary aquifers, such as the 'D' sandstone above the High Main workings at the Horden mine (Figure 3a), and by the existence of saturated zones of Coal Measures strata below primary aquifers as experienced in the Cannock Chase coalfield (Pugh, 1980).

The presence of abnormal geological structures such as large faults or other types of geological hazards will alter the path of water migration and the levels and volumes of bed separations and it will also modify the presence of incompetent rocks. They often break the sealing effect of incompetent rocks.

In addition to the experience of the water sealing effect of the incompetent rocks at the Clifton and Cotgrave mines, the working of the Shalston Group at the Royston Drift Mine in the North Yorkshire Area provides another example (Graham, 1983).

REFERENCES

Elliott, R.E. 1978. Strain zones and water migration above a working panel. Unpublished Geological Branch Seminar paper, National Coal Board, London

Farmer, I. 1984. Engineering behaviour of rocks. Chapman & Hall, London.

Garritty, P. 1981. Effects of mining of surface and subsurface water bodies. PhD thesis, University of Newcastle upon Tyne.

Garritty, P. 1982. Water percolation into fully caved longwall faces. Strata Mechanics, Ed. I.W. Farmer, Elsevier, Amsterdam, pp25-29.

Garritty, P. 1983. Water inflow into undersea mine workings. Int. Jl. Mining Engineering, Vol.1, pp237-251

Graham, S.G. 1983. Water associated with first weights on retreat faces at Royston Drift Mine below Coal Measure aquifers. Unpublished Geological Branch Seminar paper, National Coal Board, London

Hindmarsh, W.E. 1990. Asfordby - a new mine for the future. Trans. Instn. Metall. (Sect. A: Min. industry), Vol.99, A1-A14

Hoare, R.H. and Whitworth, K.R. 1978. Water occurrences affecting coal faces in the Cannock Chase coalfield, Western Area. Unpublished Geological Branch Seminar paper, National Coal Board, London

Pugh, W.L. 1980. Water problems at West Cannock No.5 colliery. Mining Engineer Vol.139, pp669-679

Wilby, G.E. 1990. The first five years of a new mine. Mining Engineer, Vol.150, pp23-28

Xiao, G.C., Irvin, R.A. and Farmer, I.W. 1991a. Use of database in ground control to identify weightinh and water inflows. Proc. tenth International Conference on Ground Control in Mining, West Virginia University, U.S.A.

Xiao, G.C., Irvin, R.A. and Farmer, I.W. 1991. Water inflows into longwall workings in the proximity of aquifer rocks. Mining Engineer, Vol.151, pp9-13

Effects of Geomechanics on Mine Design, Kidybiński & Dubiński (eds) © 1992 Balkema, Rotterdam. ISBN 90 5410 040 0

Surface subsidence associated with partial extraction: An Australian case study

D. Rowlands
Department of Mining & Metallurgical Engineering, The University of Queensland, Brisbane, Qld, Australia

ABSTRACT

This paper describes the fieldwork associated with subsidence levelling and Time Domain Reflectometry arising from a research project sponsored by the Australian National Energy Research Development and Demonstraion Programme. The project was designed to examine strata displacements associated with the partial extraction of an underground coal seam.

Scaled physical modelling with a Base Friction Rig was used to simulate and quantify the displacements. A novel method utilizing an Intergral Grid indicates how a predictive subsidence technique can be developed from base friction models.

1. INTRODUCTION

Strata displacements resulting from the underground extraction of coal seams can be considered under two classifications.

    (a)   Displacements associated with "total" seam extraction

    (b)   Displacements associated with "partial" seam extraction

Much research has been undertaken and recorded worldwide with respect to total seam extraction particularly in the U.K., Germany, Poland,and China. Strata displacements associated with partial seam extraction have not been widely researched and very little if any quality data are available for mine design purposes.

Underground coal seam extraction results in strata displacements that may extend from the seam horizon to the surface, and to a lesser extent for some distance below the seam horizon. The extent and the magnitude of these displacements, both vertical and horizontal, depend upon the dimensions of the underground excavation. If a super critical area is extracted a subsidence basin develops at the surface the geometry of which is such that a maximum value of vertical displacement occurs over a central area of the basin. Reduction of the extracted area to a critical value results in a central point on the surface exhibiting the maximum value of vertical displacement. Further reduction in the area of extraction reduces the area of the surface subsidence basin and the magnitude of the vertical displacement at the central point. This process and effect begs the basic question - Is there a threshold value of the extraction area below which the surface is not disturbed? Or in practical terms what area of a coal seam can one extract such that the associated surface displacements will cause no significant damage to surface structures? Observing the problem from the other extreme may be simpler to conceive. Single or multiple narrow entries can be driven underground with no discernable surface displacements. How wide can these entries be made before surface displacements can be observed and measured or considered significant from an engineering viewpoint?

1.1 Research Programme

This research project was aimed at developing methods of defining the strata displacements associated with partial extraction with a view to developing a design rationale.

The project was organised into the following sub-programmes.

i   Obtain diamond drill core samples to establish the physical properties of the strata in the region under consideration.

ii   Undertake small scale physical modelling using a base friction rig to obtain qualitative appreciation of the strata displacements.

iii   Develop computerized mathematical models using data obtained from sub-programmes i, ii.

iv   Ratify the mathematical models by Time Domain Reflectometry (TDR) techniques and surface subsidence surveys at Oakleigh Colliery.

1.2 The Oakleigh Colliery Programme

Oakleigh Coliery is situated some 60 kilometres west of Brisbane, the State Capital of Queensland, Australia. Coal is produced from the Walloon Coal Measures by both opencut and underground mining. The Walloon Coal Measures are of Middle Jurassic age,and at Oakleigh consist of interbedded mudstones,siltstones,and sandstones with stone-banded coalseams up to 3.5 metres thick.The coals are high volatile and used as general industry thermal fuels.

The Oakleigh Colliery work programme comprises two components:

(a)   Surface Subsidence surveying,

(b)   Time Domain Reflectometry measuring.

Four levelling grid lines were laid out at the surface over an area that had been developed by bord and pillar methods using standard continuous miners and shuttle cars, see Figure 1. The general plan of extraction was somewhat flexible, the limits of working being defined by the quality and thickness of the coal seam. Once the economic mining limit had been reached in the development roadways the

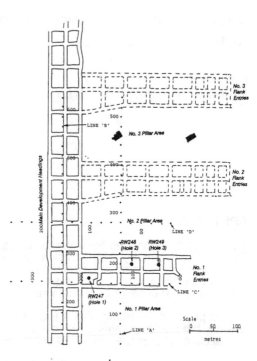

FIG. 1 Extraction Plans and Positions of Boreholes 1, 2 & 3.

pillars were systematically extracted with a general retreat outbye. Successive flank panels were planned to allow extraction of the coal on the inbye side of each set of entries.

It was realized that the geometry of the underground extraction area was not ideal from a research project viewpoint. Further, because of the existence of houses on the surface, a situation would develop where some "narrow working" would take place adjacent to a protection pillar.

It was therefore decided to divide the TDR measuring work into two phases. An initial phase using three drill holes RW247, RW248 and RW249, and a later phase using three more drillholes (4,5, and 6) to be drilled in the main development pillars adjacent to the protection pillars.

Holes RW247, RW248 and RW249 were drilled by the Drilling Branch staff, Queensland Department of Mines. Having no previous experience of grouting cables into drillholes it was decided to practice by grouting one cable into hole RW247, then having perfected the method, grout two cables into each of the other holes so that check measurements could be made.

28

The success of the TDR measuring programme depended upon a perfect bond between the coaxial cable and the side walls of the drill hole. It was therefore decided that an expansive additive (DENKA CSA) should be mixed with the cement water grout to achieve this result. Trials, using split steel tube moulds, were carried out to confirm the expansive effect, followed by pull out tests to prove the adhesive forces exceeded the ultimate breaking tensile force for the coaxial cable. The grout to cable bond was increased by the presence of metal hose clips crimped onto the cable at specified intervals. These crimps provided a reflected signal "signature" which facilitated interpretation of the TDR recordings.

Hole RW247 was diamond drilled and cored to provide a suite of samples for physical testing. Geophysical logs were obtained for all three holes and a geological description of the core was provided by staff of the Queensland Geological Survey Department.

Uniaxial compressive strength tests were carried out on the drill core samples to provide the basis for scaling material properties for base friction modelling. Tensile strengths were obtained from rock discs using the indirect Brazilian test, and the elastic modulus and Poison's ratio were measured for use in mathematical modelling by the Finite Element technique.

2. DETAILED DESCRIPTIONS OF SUB PROGRAMMES

2.1 Surface Levelling at Oakleigh Colliery

The surface levelling operations were undertaken by the Surveying Branch of the Queensland Mines Department. The initial survey was made on 15.12.87 with periodic measurement continuing to the end of 1989. Stations were laid out at 20 metre intervals along two lines running parallel with the main development headings (A and B) and two lines (C and D) parallel with the panel headings.

Details of the progressive vertical displacements of the levelling stations are illustrated in Figures 2, 3, 4 and 5. With respect to Figure 2, the diagrammatic shape of the surface trough between stations A80 and A180 is obviously

incorrect. The two points in the figure have been joined by a straight line because of the absence of levelling data. If data were available we would expect a trough similar to that between stations A280 and A360 as shown to the right in Figure 2. A similar comment applies to the straight line shown in the figure between stations A400 and A500. In this case one would expect displacements and surface curvature similar to that shown between stations A180 and A280 as of 31.08.88. The "ridge" between stations A180 and A280 is associated with the partial extraction of coal pillars.

Line B, to the west of line A, was positioned directly above the main development headings bearing $8^o 4' 40"$. All stations were accessible hence there were no problems with missing data as previously mentioned above with Line A.

Line C was positioned above and parallel with the 1st Panel headings bearing $98^o 40'40"$. The absence of levelling data at Stations C60 and C260 does not appear to have a major effect upon the surface subsidence profile. Line D is parallel to Line C and 120 metres North.

2.2 Time Domain Reflectometry Measurement at Oakleigh Colliery

The principles involved in Time Domain Reflectometry (TDR) as applied to measurements of rock deformation have been very well described by Dowding, Su and O'Connors (1988). As they state the technique was originally developed to locate faults in coaxial cables and had been used by Panek and Tesch (1981). In essence the technique utilizes the reflection of an electric pulse from deformations in a coaxial cable. The travel time of the reflected pulse is related to the distance of the deformation from the instrument source, and the shape and orientation of the pulse indicates the type of deformation involved. The instrument is simple to operate and does not require a knowledge of the fundamentals involved.

The initial TDR programme associated with the coaxial cables grouted in holes RW247, RW 248 and RW 249 was designed to obtain the maximum number of measurements during the most active period of overburden movement. When the underground extraction commenced in the vicinity of hole RW 249 readings were

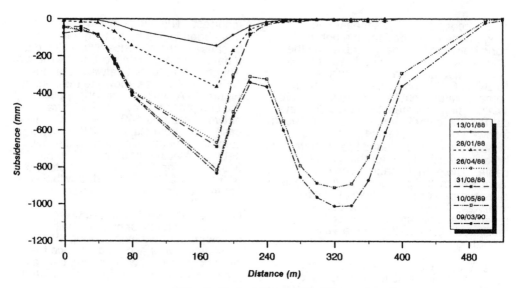

FIG. 2 Subsidence Profiles Line A.

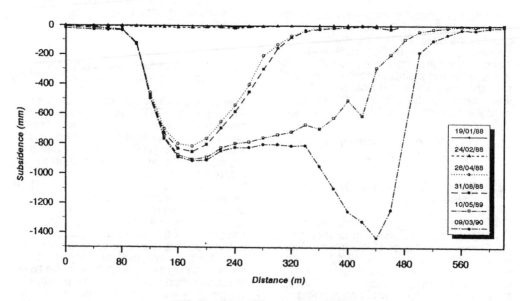

FIG. 3 Subsidence Profiles Line B.

taken weekly. As the area of extraction increased and the working faces moved away from the vicinity of the holes, the measurements were taken monthly.

As previously mentioned miniature hose clips were crimped onto the coaxial cable to provide a reference scale for reflection measurements.

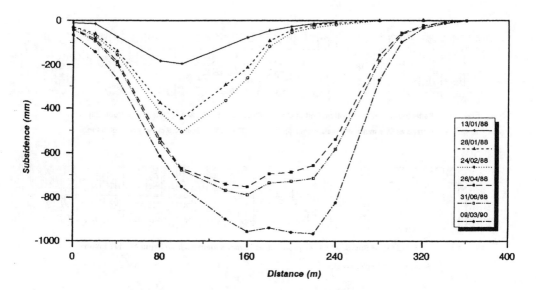

FIG. 4 Subsidence Profiles Line C.

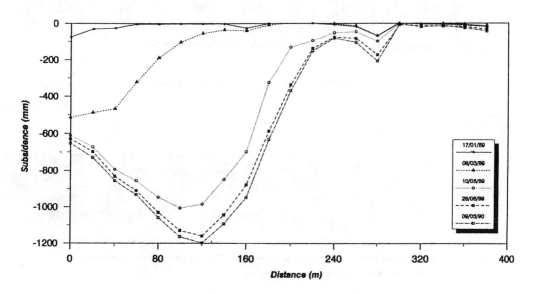

FIG. 5 Subsidence Profiles Line D.

The coaxial cables were high quality, low attenuation, solid copper core (BELDEN 9914 GG 5.26 sq mm Cu core 2.31 mm foam PE dilectric) supplied in 303 metre lengths. Ten hose clips were crimped at one metre intervals to the lower ends of the cables to monitor the caving of the roof in the goaf area. The remainder of the cables were crimped at 10 metre intervals. Typical signals from an undisturbed cable in hole RW 249 are illustrated in Figures 6a, b & c.

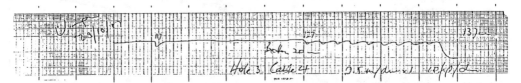

FIG. 6a   Reflected Signal from Bottom End of Coaxial Cable 4, Hole RW249.   Note the 10 Hose Clip Crimps at One metre intervals from Depth 127 - 137 metres. Scale 2.5 m per major division.

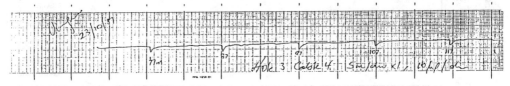

FIG. 6b   Reflected Signal Showing Hose Clip Crimps at 10 metre intervals from Depth 77 - 117 metres. Scale 5 m per major division.

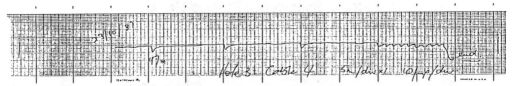

FIG. 6c   Reflected Signal from Bottom End of Coaxial Cable.   Scale 5 m per major division.

FIGURES 6a, b & c (Prior to Mining) 23/10/87

FIG. 7a   Reflected Signal from Bottom End of Coaxial Cable 4, Hole RW249. Note that 8 metres of Crimped Cable have been Removed due to Roof Caving. Two Extra Crimps have Developed due to Shear Displacement at 111 m and 118.5 m. Scale 2.5 m per major division.

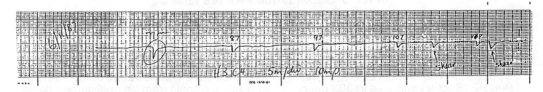

FIG. 7b   Reflected Signal from 77 m - 117 m Showing Shear Crimps.   Scale 5 m per major division.

FIGURES 7a & b (Pillar Extraction In Progress) 6/11/87

A special jig was manufactured to ensure that all the hose clips were crimped onto the coaxial cable in a like manner, i.e. to ensure the geometry of the cable deformation was constant and thereby produce an identical reflection.

TDR readings were taken from the cables before pillar extraction commenced and immediately after the first pillars were extracted. The initial caving was indicated by a shortening of the two cables in hole RW 249. Further caving was noted as the second pillar was extracted.

The recorded traces obtained from the two cables before and after extraction commenced are shown in Figures 7a and b. After the first pillar was extracted the roof caved above the seam for some 2.0m. This increased to 8.2m after the second pillar was extracted. These TDR measurement were confirmed by underground observations of the caved area.

In addition to indicating the loss of cable at the bottom of the hole two extra negative reflections are shown at 118.5m and 111.25m. These reflections are due to shear deformations of the cable resulting from roof strata displacements.

TDR results for the cables grouted into the initial three bore holes are illustrated in Figures 8a, b, c. Each dash indicates a reflection from a hose clip or cable deformation caused by strata displacements.

Note that cable 1 was grouted into hole RW 247
  Cables 2 & 3 were grouted into hole RW 248
  Cables 4 & 5 were grouted into hole RW 249.

The positions of negative reflections from hose clips and shear deformations are shown as measured vertically from the collar of the bore hole in metres. They have been plotted against a time base measured in days from the commencement of measurements on 10/11/87.

Interpretation of these results has to be carried out in association with the underground pillar extraction programme, to obtain the maximum information. This can be a very time consuming exercise requiring much patience. Nevertheless, some important details can be obtained rapidly as described in the following.

Examination of the results for Cable 1 shows that the overburden was not disturbed until day 50 (23/12/87) when a shear displacement developed some 48 metres below the hole collar or 86 metres above the seam roof. (It is simpler to configure the geometry by measuring from the seam horizon because it eliminates problems with surface topography). By this time the goaf edge had advanced to a position 60m from the bottom of hole RW 247 indicating movement on a plane angled from the vertical at 35°. In addition, as the extraction face progressed towards hole 247 additional shear displacements were indicated predominantly less than 86m above the seam some 80 days after measurements commenced. These measurements are consistent with the concept of an advancing inclined plane hading upwards from the goaf line. By the 80th day (21/01/88) a shear crimp is indicated 69 metres above the seam horizon at which time the goaf line was 40-m away, giving an angle of 30°. Further, on the 73rd day (14/01/88) the lower section of the coaxial cable was severed 54m above the seam when the extraction line was between 60m and 86m away. Taking the mean of these two distances at 73m this corresponds to an inclined plane hading at 36° upwards. The final loss of cable on the 25/02/88 i.e. the 115 day was complicated by the second extraction face advancing at right angles from the Southern end of the main development drives. Surface subsidence measurements taken on the 24/02/88 show that Stns 180 and 200 on line C adjacent to Hole RW 247 had vertical displacements of 117mm and 54mm respectively, being in the zone of maximum surface tension. All of the above discussions would indicate an "angle of draw" of some 35° to 36° from the vertical for the Western end of Line C.

The pattern of induced shears and cable losses for cables 2 and 3 in Hole RW 248 from day 38 (11/12/87) indicate that as soon as caving commenced i.e. a loss of some 18m from the lower end of the cable, the overburden tended to sag into a trough. This caused shear displacements between adjacent strata due to their varying curvatures, see Figure 9. The end of the cable was progressively severed as the shear displacement increased.

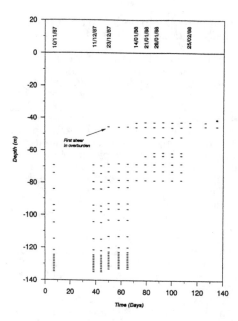

FIG. 8a   TDR Signal Summary Hole RW247 Cable 1.

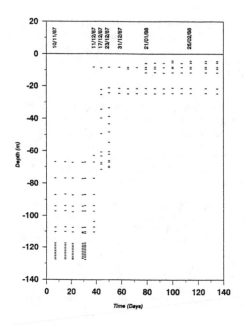

FIG. 8b   TDR Signal Summary Hole RW248 Cable 2.

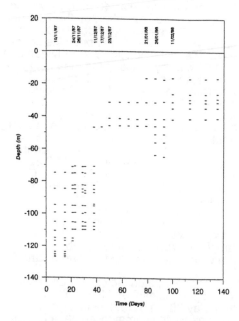

FIG. 8c   TDR Signal Summary Hole RW249 Cable 4.

The pattern of induced shears and cable losses for cables 4 and 5 in hole RW 249 are again different. It must be remembered that this hole was placed over the first pillar near the coal rib from which extraction commenced. After a certain area of seam was extracted the roof started to cave and there was a loss of cable. Extraction thereafter would progress away from and to one side of the hole thereby inducing shear deformations at increasingly higher horizons as the area of extraction increased, see Figure 10.

2.3   Base Friction Physical Modelling

An extensive analysis of the fundamental principles involved in base friction modelling has been carried out by Bray & Goodman (1981). They showed that as stress increases with depth of superincumbent strata in the real prototype situation, so also does stress increase with depth in the model situation.

Consider a simple base friction model as illustrated in Figure 11, i.e. a cohesive sandy model material of uniform thickness (t) and dimensions (a) and (b) placed upon an endless belt. The model material is prevented from moving when the belt is driven in the direction indicated by the restraining bar. The restraining force depends upon the weight of the material upon the belt and the coefficient of friction between the belt and the model.

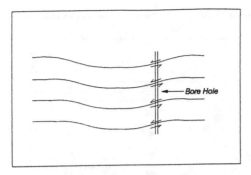

FIG. 9   Shear Displacement in a hole passing through strata
          forming a subsidence trough

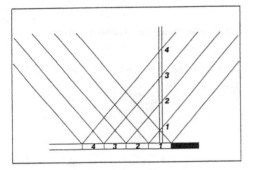

FIG. 10   Successive Shears in bore hole as extraction
           extends from Right to Left

For a fixed width (a) and thickness (t) the restraining force increases directly with dimension (b). In the model the dimension (b) is analogous to depth in the real situation.

If the model material has density ρ
  The mass of material = $abt\rho$
  The weight of material = $abt\rho g$
  g = acceleration due to gravity
  The restraining force = $\mu abt\rho g$
  μ = coefficient of friction between belt and model material
  The stress on the restraining bar (stress in model at depth b)
  σ_b = $\mu abt\rho g/at$
  σ_b = $\mu b\rho g$ (1)

The stress at depth z in the real situation is calculated as follows:

  σ_z = (Force/unit area)
       = z (Mass/unit volume) x g/unit area
  σz = $z\rho g$ (2)

If b = z/150 i.e. a scaling factor of 1:150
Then σ_b = $\mu\rho gz/150$
and $\sigma b/\sigma z$ = $\mu z\rho g/150z\rho g$
   σ_b = $\mu\sigma_z/150$
i.e., σ_b is proportional to $\sigma_z/150$ (3)

The model stress is scaled in direct proportion to the geometric scaling factor. Note the above relationship assumes that the densities of the model and real materials are the same.

2.3.1 Modelling the Stratigraphic Sequence at Oakleigh Colliery

The geometric scaling factor depends upon the depth of workings and the size of the modelling rig. The depths of workings at Oakleigh were between 130m and 140m below the surface, and dimension (b) in the model was approximately 1m, hence a geometric scaling factor of 1:150 was chosen.

The geological description of the stratigraphic sequence down to the Cowell Seam from Bore hole RW 247 Oakleigh Colliery is given in Table 1, and the corresponding geomechanical properties are given in Table 2.

Missing values in Table 2 indicate materials that were too weak to test and are considered below.

In order to simplify the physical modelling it was decided to divide the sequence into five compressive strength ranges.

  Range 1   60 - 62 MPa
  Range 2   27 - 35 MPa
  Range 3   13 - 23 MPa
  Range 4   8 - 16 MPa
  Range 5   <2   MPa

Strata layers 7 and 9, 16, 17 and 18 were placed in range 4 and the weathered basalt material 19 at the surface was placed in range 5.

2.3.2 Modelling Materials

After a review of the literature and a series of laboratory tests it was decided to use a modelling material based upon fine sand, coarse sand, builder's plaster, and water. The material is easy to handle and clean up, is cheap, has elasto-plastic properties, but has markedly time dependent strength properties. This latter fact imposes a time

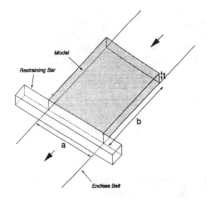

FIG. 11 Physical Conditions, Base Friction Modelling.

(Model, Restraining Bar, Endless Belt, a, b)

TABLE 2 : Geomechanical Properties of the Stratigraphic Sequence

Strata Layer	Compressive Strength (Mpa)	Young's Modulus (GPs)
19	-	-
18	-	-
17	-	-
16	-	-
15	18.3	2.5
14	61.7	22.1
13	16.6	1.3
12	17.3	2.0
11	22.8	3.3
10	18.8	3.5
9	-	-
8	34.5	7.1
7	-	-
6	27.8	11.0
5	33.0	5.2
4	34.1	7.7
3	13.0	-
2	28.4	8.2
1	27.0	8.6

TABLE 1 : Geological Description of the Stratigraphic Sequence

Strata Layer	Depth (m)	Thickness (m)	Geological Description
19	0.0 - 26.7	26.7	Highly weathered Basalt, Claystone and Sandstone
18	26.7 - 43.5	16.8	Grey Mudstones
17	43.5 - 49.7	6.2	Mudstone and coal laminated structure
16	49.7 - 54.5	4.8	Light grey Siltstones grading into Mudstones
15	54.5 - 66.4	11.9	Sandstones containing mineral and organic fragments
14	66.4 - 66.7	2.3	Homogeneous Sandstone band
13	66.7 - 72.5	3.8	Mudstone containing silty FeCo modules in parts
12	72.5 - 76.7	4.2	Light grey Sandstone containing Siltstone beds
11	76.7 - 80.8	4.1	Irregularly banded Sandstones Mudstones and coaly beds
10	80.8 - 82.3	1.5	Mudstone with little Limestone infil
9	82.3 - 93.6	11.3	Grey Mudstone and Siltstone laminations
8	93.6 - 95.5	1.9	Light grey Sandstone layer
7	95.5 - 99.8	4.3	Siltstone and Mudstone laminations
6	99.8 - 112.5	12.7	Sandstone bed containing Siltstone and elastic sediments
5	112.5 - 119.2	16.7	Interbedded mudstone, Sandstone and Coal
4	119.2 - 130.0	10.8	Uniform Sandstone layer with a few Siltstone bands
3	130.0 - 133.6	3.6	Banded coal seam - Cowall seam
2	133.6 - 147.9	14.3	Interbedded Sandstone and Mudstone layers
1	147.9 - 150.0	2.1	Sandstone

TABLE 3 : Model Material Compositions and Strengths

Material No.	Mixture	Strength (kPa)
1	7% Plaster 15% Water 78% Fine Sand	205
2	4% Plaster 12% Water 42% Fine Sand 42% Coarse Sand	110
3	4% Plaster 14% Water 82% Fine Sand	55
4	3.5% Plaster 12% Water 84.5% Fine Sand	40
5	5% Water 95% Fine Sand	-

compactor mounted on the frame of the base friction rig.

The mixtures and strengths of the five ranges of model materials are listed in Table 3.

constraint when model testing. As will be shown later a typical physical model could be made up ready for testing in two hours, thus the model material strengths were measured two hours after casting in cylindrical moulds 38 mm diameter.

The dry ingredients, sand and plaster, were thoroughly mixed before adding water. The moist material was then compacted in the mould using three equal volumes at a standard pressure of 7 kPa. This standard value is obtained from the weight and dimensions of the travelling

2.3.3 Base Friction Rig and Measuring Instrument

The essential components of the base friction rig are as follows and shown in Figure 12:

1. Frame
2. Endless Belt and Rollers
3. Hand Drive
4. Travelling Compaction Device
5. Glass Cover Sheet
6. Displacement measuring instrument

Travelling
Compaction
Device

Frame

Hand Drive

Endless Belt

Roller

FIG. 12 **Base Friction Rig.**

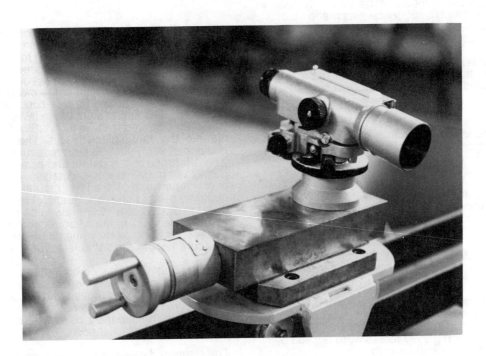

FIG. 13 **Tilting Level, Micrometer, and Stand.**

Displacements of the model materials in the direction of, and normal to, the travel of the belt were made with an engineers tilting level bolted to a compound tool feed mount off a metal working lathe. Direct reading micrometers allowed displacements down to 0.0254 mm to be resolved, which is equivalent of levelling to an accuracy of 4 mm with a 1:150 scaling factor. A new technique has recently been perfected using optical scales and digital readouts which can resolve displacements of 0.001mm.

A channel iron section mounted upon a four legged stand with adjustable screw feet is used as the carriage guide for the tilting level. This stand allows the level to travel parallel to the rig distanced some 1 m away. The optics of the level telescope allow accurate observations at such distances, Figure 13.

2.3.4 Making and Testing Models

A simplified model of the Oakleigh Colliery stratigraphic sequence was made up of the nineteen layers listed in Table 2 and described in Section 2.3.1.

Modelled bedding planes were then cut into the various layers by running a spatula through the material using an angle iron as a guide.

Joints were also cut at the appropriate attitude using a narrow spatula. Joint spacing and direction were based upon underground observations of the roof strata at Oakleigh Colliery.

The armoured glass cover sheet was then clamped to the top frame and the belt wound forward to ensure all bedding planes were closed and compacted.

The glass sheet was then removed and the scaled excavation was cut in the model.

Displacements were measured by observing the positions of reference pins placed in the model before and after the simulated gravity forces were applied by driving the belt forward.

Reference pins with coloured heads were inserted into the model at three horizons from a datum 13.5 cms below the seam roof i.e. 30 cm, 60 cm and 90 cm from the frame front member. The pins were inserted vertically using a grooved

template at 50 mm spacings i.e. equivalent to 7.5 metres at Oakleigh. The pins were placed symmetrical each side of a line passing through the centre of the excavation, and pressed down level with the model surface, Figure 14.

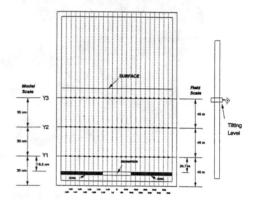

FIG. 14 Reference Pins at Three Horizons in the Model.

A datum was established at each horizon by placing a tensioned fine black thread across the model between scribed marks on the side members of the top frame. The tilting level stand was then set up parallel with the side frame of the model such that both near and far side could be focussed. The line of sight of the level was then adjusted to be in the vertical plane of the datum black thread. The position of each pin was then measured with reference to this datum and assigned positive or negative y co-ordinates. The micrometer head on the level mounting could be read to one thousandth of an inch i.e. 0.0254 mm. Care had to be taken to eliminate slack in the micrometer feed screw by making all movements before a measurement in the same direction.

After establishing the y co-ordinates for all pins at each of the three horizons, the tilting level stand was moved and set up parallel with the front drive roller and using the same technique the x co-ordinates of each pin were established.

The armoured cover glass plate was then replaced and clamped over the model, prior to testing.

Testing consisted of slowly winding the conveyor belt forward under the model at a uniform speed and observing the displacements of the modelled strata.

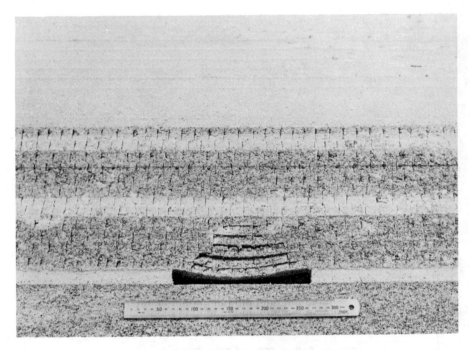

FIG. 15 Failure Pattern, 30 m Excavation.

FIG. 16 Failure Pattern, 40 m Excavation.

FIG. 17 **Failure Pattern, 50 m Excavation.**

After some initial tests it was realized that the movement of the belt resulted in some material being abraided from the base of the model. This meant that a correction had to be made to all measured results, depending upon the distance moved by the belt.A simple experiment was set up to establish the effect of abrasion on the pin movement without an excavation in the model.

The patterns of failure for 30m, 40m and 50m openings are shown in Figures 15, 16, and 17. The limits of fracture are obviously controlled by the joints and bedding planes. The lack of symmetry about the centre of the excavation arose from the fact that the joint planes were not normal to the plane of the belt, but plunging from right to left in the figures. Generally the span of the failure zone decreased with distance from the excavation as did the degree of compaction. Bed separation without major failure can be observed together with some outer fibre tensile failure initiating from the upper surfaces of some beds. The failure patterns are typical of those observed from the edge of the goaf and sometimes exposed in opencut mines that intersect old workings.

It should be noted that horizon Y3 was nearest to the surface and indicated the near surface displacements.

With a typical angle of draw of 35° from the vertical, and a working depth of 135 metres, the critical area of extraction for a point on the surface would have a diameter D where

$$D = 2(135 \tan 35^\circ)m$$
$$= 189m$$

The 50m wide excavation was therefore only 26% of the critical diameter. Further discussions with respect to these results are contained in Section 4.

3. INFLUENCE FUNCTION AND INTEGRATION GRIDS

An English translation of the review text by Helmut Kratzsch titled "Mining Subsidence Engineering" was published by Springer-Verlag in 1983. Much of that text is devoted to the European approach to subsidence engineering which has recently been adopted by the Chinese.

The techniques employ both empirical and theoretical concepts and are used extensively in Germany and Poland in mine design and planning. The empirical components involve back analysis to supply basic information concerning the geometry of subsidence basins. This information is then incorporated into theoretical models to predict future subsidence. Obviously the predictions tend to be site or regionally specific because of the back analysis components involved. Nevertheless, once good quality data has been established the techniques are very useful.

The basic concepts involved in developing an influence function are illustrated in Figure 18. If a critical sized element of coal is removed from a near horizontal coal seam, then the overlying and near excavation strata are displaced as indicated by the size and direction of the vectors.

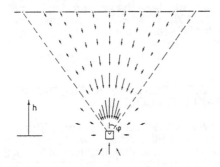

FIG. 18   **Movement Vectors Caused by an Element of Excavation**   (After Kratzsch).

FIG. 18   **Movement Vectors Caused by an**

Element of Excavation   (After Kratzsch).

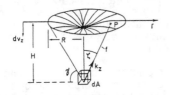

FIG. 19   **The Elementary Trough Created at the Surface**

by the Extraction of Element dA   (After Kratzsch).

Considering the surface displacements, the maximum values occur centrally above the excavation and they diminish radially to zero at the limit angle φ. The excavation has a much greater influence upon the surface displacements directly above than it has upon points some distance away.

The three dimensional implications of this concept are illustrated in Figure 19. The influence k_z of excavating the element dA on the point P at the surface can be expressed as a funciton of the radial distance r, the inclined distance f, or the subtended vertical angle ζ.

Thus K_z can be expressed as $K_z(r)$, $K_z(f)$, $K_z(\zeta)$. The shape of this influence function obviously depends upon the physical properties of the material above the excavation, and tends to be site or regionally specific.

Where the overburden material is grannular and non cohesive the influence function will be centrally peaked. At the other extreme where the overburden is cohesive and elastic the influence function will be far less peaked at the centre and more widely distributed. For elasto plastic conditions the influence function will be intermediate between these two extremes.

For design purposes it is convenient to convert the influence of mining individual elements into an integral grid such that the cumulative effect of mining finite areas can be calculated. This is illustrated in Figure 20.

Considering the surface point P, the extraction of central element dA_2 at depth H has a much greater influence on P than does the extraction of element dA_1, at radius r. The actual influence k_z of any element is given by the vertical ordinate of that element on the influence function curve. The value of k_z can be established by considering the three dimensional implications of Figure 20. The critical area of extraction at depth H for point P is given by the circle of radius R where $R = H \cot \gamma$. If we consider the volume of revolution of the influence function curve as e then $k_z = de/dA$, where de is the cylindrical volume under the three dimensional surface of the influence function delineated by the circular strip of width dA.

If all the critical area πR^2 is extracted then P will subside by a maximum amount V_z max. If some lesser area within the critical

41

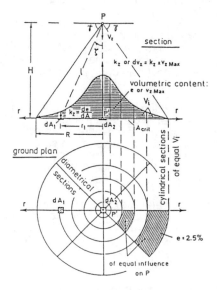

FIG. 20 Construction of an Integration Grid

(After Kratzsch).

concentric circles and radial lines is constructed on transparent material at the appropriate scale so that it can be moved systematically over working plans to estimate the surface displacements. Calculations involve simple arithmetic and visual estimates of areas enclosed by the circular grid segments. If all the critical area of influence is extracted then the surface point will be subjected to maximum vertical displacement. If the area of extraction is equivalent to 20 circular segments of the critical circular area, then the subsidence will be 20 x 2.5 = 50% of maximum vertical displacement.

As reported by Kratzsch (page 237), Brauner (1961) developed an equation to calculate the radii of circles used in an integral grid for a cohesionless model medium (stochastic)

In his equation

$$r = R \left\{ 1/\pi [\ln 1/(1 - V_z)] \right\}^{0.5} \quad (5)$$

where the value r corresponds with V_z which may take the accepted values of 20%, 40%, 60%, 80% of V_z max. The outer fifth zone calculated to have an infinite radius from the above expression can be limited to R with no practical error.

radius R is extracted, then P will subside by a lesser amount dependent upon the area extracted and its position relative to the central line through P.

If the shape of the influence function is known it is possible to construct a series of concentric circles such that areas delineated between successive circles have equal influence upon the subsidence of point P.

Historically the critical area has been subdivided into five concentric areas of equal influence (20%), and each of these areas has been further subdivided into eight radial segments each of which has an influence of 2.5%.

A number of integral grids have been calculated by various researches as indicated in Table 4 (after Kratzsch).

The integral grids are used to calculate the vertical surface displacements resulting from extraction within the critical areas subtended by the limit angles of draw for a point on the surface. Consider the situation where a horizontal seam is being mined, and the limit angle of draw (measured from the horizontal is 55°) the critical mining area will be a circle of radius R = H Cot 55°. The integral grid of

4. DISCUSSION OF RESULTS FROM OAKLEIGH COLLIERY

By August 1989 all extraction in Panels 1 and 2 and Pillar Areas 1 and 2 had been completed, and development of Panel 3 had commenced. The extracted areas were as shown in Figure 21.

Survey Line D is shown to be above Pillar Area 2, the eastern end of which was only partially extracted. It was therefore decided to use the subsidence data from line D to assess the utility of integral grids. Utilizing the data from Table 4 and Eqn (5) three integral grids were drawn to scale with the critical outer radius R based upon a limit angle of 55° measured from the horizontal.

The percentage extraction was calculated for surface points represented by every second survey station along line D starting from station 00. Thus data was obtained for positions 40 metres apart along Line D. The estimated percentage extractions ranged from approximately 50% in the East to 90% near the centre between Lines A & B.

TABLE 4 : Subsidence Grid Radii as a percentage of the Critical Area Radius R to form five concentric circles of equal influence on a central point at the surface.

Zone (from the centre outwards)	Integration Grid			
	Bals	Stochastic	Sann	Ehrhardt Sauer
1	0.16	0.26	0.09	0.22
2	0.30	0.40	0.18	0.33
3	0.49	0.54	0.30	0.45
4	0.70	0.71	0.45	0.59
5	1.00	1.00	1.00	1.00

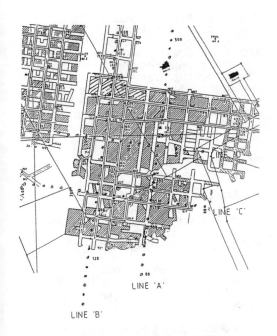

LINE 'C'

LINE 'A'

LINE 'B'

FIG. 21 Cowell Seam Extraction, August 1989.

In order to obtain data for less than 50% extraction the exercise was extended to include Line C starting from the Western end Stn 340. The results are shown in Figure 22 for the three grids, Bals, Sann and the stochastic grid of Equation (5).

Linear regression analysis for each grid shows the data for Sann to be more scattered with the R^2 coefficient being 0.892, and the zero displacement on the x axis at 10.9% extraction. For Bals' and the stochastic grid the R^2 coefficients are 0.960 and 0.981 respectively with the zero displacements at 20 and 20.3 percent.

These findings are very signficant because they indicate a method of answering the original question posed in the Introduction to the project, namely

"How wide can single or multiple entries be made before surface displacements can be observed or measured". The very high degree of correlation shown by the Bals' and stochastic grids indicate that the answer is approximately 20% of the critical area.

Bearing in mind that the diameter of the critical area of influence for a point on the surface at Oakleigh Colliery is some 189 metres then 20% may be equated to a single panel of width 38 metres. Excavations exceeding this size would resuilt in measurable surface subsidence. This then indicates a method of utilizing the data obtained from the base friction model tests.

Having established the minimum size of opening that will cause measurable surface displacements it is then possible to model a slightly larger excavation to obtain an "elementary" surface subsidence trough. The trough produced by any larger excavation can then be obtained by summing a number of elementary troughs resulting from a series of adjacent excavations. As an example using the Oakleigh Colliery results for a 50 metre wide model excavation, we can develp a "composite" trough for a 200 metre wide excavation by summing four geometrically spaced elementary troughs resulting from four adjacent 50 metre excavations. Figure 23 illustrates this procedure using the average displacements from two 50 metre model tests to produce an elementary trough, four of which were geometrically summed to give the composite trough that would result from a

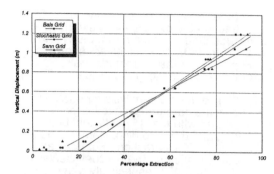

FIG. 22 Critical Area Analysis along Line C and Line D.

200 metre wide excavation.

A comparison of the model results with actual field measurements is shown in Figure 24 where the composite modelled trough has been superimposed upon the actual trough measured alone Line C. The comparison is very good and indicates that the base friction modelling technique has the potential to become a very useful method of predicting subsidence. Obviously more tests using data from a number of mine sites are necessary to prove the system.

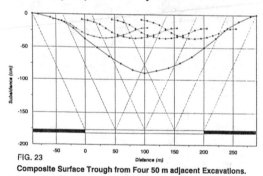

FIG. 23
Composite Surface Trough from Four 50 m adjacent Excavations.

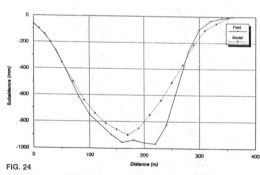

FIG. 24
Composite Surface Trough Based Upon 50 m Model Troughs Superimposed Upon Measured Trough along Line C, Oakleigh Colliery.

5. SUMMARY AND CONCLUSIONS

Base friction modelling is potentially a technique that can produce quantitative results to predict subsidence. The elementary subsidence troughs measured from modelled excavations can be added geometrically to predict the shape of the subsidence trough produced by a larger excavation.

The results from applying influence grids to surface subsidence indicate that if more than 20% of the critical area of a coal seam is extracted, then some surface subsidence can be expected.

Time domain reflectrometry can provide some information about the behaviour of insitu rock masses as they are being undermined. The technique is not cheap when one considers the total costs of drilling, grouting, measuring and recording, plus equipment and consumables. TDR techniques are very sensitive to shear displacement, but once the cable is sheared it is not possible to obtain measurements beyond the point of shear.

The knowledge obtained from base friction modelling has indicated where improvements can be made to the equipment and operating techniques to speed up measuring, increase the accuracy of displacement measuring, and obtain load measurements from the coal ribs. These improvements and modifications are currently underway, financed by a second stage NERDDC grant. The second stage of the project will use data from another two underground mines in Queensland and also demonstrate the application of the technique to surface mining.

6. ACKNOWLEDGEMENTS

The author wishes to acknowledge the various contributions made by the following people during this project.

Dr.Christopher Meimaris and Mr.Sivakumar Theivendrampillai for their invaluable services as Senior Research Officers.

Mr.Andrew Rosengren for laboratory testing.

The workshop staff, Department of Mining & Metallurgical Engineering for the manufacture of the base friction rig.

The Management and staff of the
Oakleigh Colliery, Rosewood,
Queensland, for assistance in fieldwork.

The Department of Mines, Queensland for
services rendered in diamond drilling and
surface subsidence surveying.

7. REFERENCES

Bray, J.W., Goodman, R.E., (1981) "The
Theory of Base Friction Models", Int. J.
Rock Mech. Min. Sci., Vol.18, pp.453-468.

Dowding, C.H., Su, M.B., O'Connors, K.,
(1988) "Principles of Time Domain
Reflectometry Applied to Measurements of
Rock Mass Deformation", Int. J. Rock
Mech. Min. Sci., Vol.25, pp.287-297.

Kratzsch, H., (1983) Mining Subsidence
Engineering, Springer-Verlag. (Translated
by R.F.S. Fleming)

Panek, L.A., Tesch, W.J., (1981).
Monitoring Ground Movements near
Caving Slopes - methods and
measurements, U.S.B.M. R.I. 8585

Wilson, A.H., (1980) "The Stability of
Underground Workings in Soft Rocks of
the Coal Measures", Ph.D. Thesis,
University of Nottingham, U.K.

Effects of Geomechanics on Mine Design, Kidybiński & Dubiński (eds) © 1992 Balkema, Rotterdam. ISBN 90 5410 040 0

Control of deformations in the pillar between the twin bores of a tunnel in Aosta valley, Italy

P.Grasso, G.Carrieri & A.Mahtab
GEODATA, Torino, Italy

ABSTRACT: About one-third of the 2.7-km long, twin-bore tunnel discussed here traverses a morainic deposit. The ground control problems anticipated in the design and excavation of the tunnel in this weak material are discussed together with the procedures used for controlling the excessive deformations in the narrow pillar between the two bores. The two recommendations for continuing the excavation are to increase the strength of the pillar (widening and/or reinforcement) and to keep the face of the second bore behind the invert of the first bore.

1 INTRODUCTION

The tunnel discussed here, which will be henceforth referred to as the study tunnel, is one of several tunnels that form parts of the new motorway which connects Aosta in northwestern Italy to the French border near Monte Bianco (Figure 1). The 2.7-km long tunnel has two bores, each having a horse-shoe shaped cross section of about 100 m². The left and right bores are referred to the view

from the Monte Bianco portal.

The tunnel lies from 70m to 185m below the ground surface. However, the cover thickness over about one third of the tunnel adjacent to the Monte Bianco portal ranges from 70m to 90m. This is the section of the tunnel that is of interest here. The material in this portion consists of a morainic deposit which is the source of several problems of construction and stability.

In the morainic deposit near the Monte

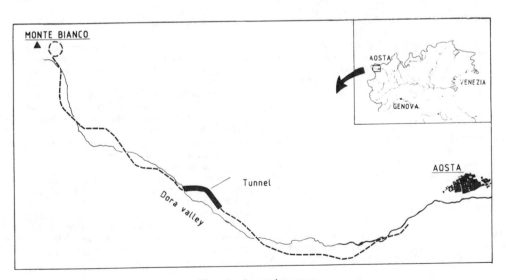

Fig. 1 Location map

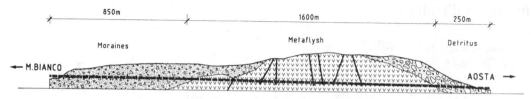

TUNNEL

850m | 1600m | 250m

Moraines | Metaflysh | Detritus

← M.BIANCO | AOSTA →

Fig. 2 Geologic section along the tunnel

Bianco portal, tunnel excavation is performed by a back hoe. The remainder of the study tunnel (toward the Aosta portal) is in a stronger formation (metaflysh) and is excavated largely by means of the conventional, drilling and blasting technique.

Excavation of the tunnel began in March 1990. Progress to the end of June 1991 consisted of an advance of about 1km from the Aosta portal and about 500m from the Monte Bianco portal.

The problem of large deformations and high requirements of support in the moraine were anticipated in the design. However, these problems were magnified by the specified, small dimension (8m) of the pillar between the two, 13-m wide bores. The objectives of the ground control measures (described in subsequent sections of this paper) are to stop the convergence of the tunnel bores and the deformation of the pillar with consequent reduction in the load on the supports.

2 GEOMECHANICAL CHARACTERIZATION

A geologic section along the tunnel axis is shown in Figure 2. Proceeding from the Monte Bianco portal toward the Aosta portal, the following formations are encountered by the tunnel:

la Salle moraines: 850m
metaflysch   : 1600m
detritus   : 250m

The problematic material for this case study is the moraine which consists of an incoherent mixture of particles of various sizes (often as boulders of volume exceeding a m^3) together with lenses of sand and sandy clay. The sketch of a tunnel face in the moraine is shown in Figure 3(a) while a picture of another typical face is shown in Figure 3(b).

The shear strength and deformation modulus for the moraine were initially selected as a result of the literature survey and experience in other construction sites in morainic deposits. These initial estimates were later revised into design values based on back-analysis using the observed deformations and loads. Both the initial and the design values are listed below.

Geotechnical parameter	Initial estimate	Design value	
		Peak	residual
angle of friction (deg.)	30-36	35	30
cohesion (kPa)	10-100	25	0
Modulus of deformation (MPa)	200-350	100	15

3 DESIGN CONCEPTS

The choice of the excavation procedures and ground control techniques to be used during construction of the Monte Bianco side of the study tunnel was dictated by the characteristics of the moraine. The low values of cohesion and the angle of friction indicated that the ground has little self-supporting capacity and that preventive consolidation or reinforcement of the ground would be necessary to obtain a stable excavation.

Moreover, the mechanical properties of the ground will weaken further in the presence of clay and water. In fact, under these conditions, the angle of friction of the material may decrease and the excavation become totally unstable at the face and around the periphery of the tunnel.

LEGEND

⬭ₒ	Boulders and pebbles
⦂⦂	Gravel and sand
≈ ≈	Silt

Morainic deposits

FACE

1° Work level

1m

Fig. 3 (a)   Sketch of face of left bore of the tunnel in moraine

Fig. 3 (b)   Photograph of face in moraine in close proximity
of the location of Fig. 3 (a)

The design concept for excavating the tunnel requires pre-reinforcement of the ground from the periphery of the opening and in advance of the face following the NATM principles. From among several techniques, the pipe-umbrella arch system was selected for the prereinforcement, using valved pipes and cement injection at low pressure.

The umbrella arch is constructed around the periphery of the heading as a subhorizontally-inclined semi cone of steel pipes installed in boreholes by a special two-arm drilling machine, and grouting through the pipes, which have valves or grout holes along the length of the pipe. This produces a uniformly consolidated rock mass between any two adjacent pipes.

A: Subhorizontal umbrella pipes;

B: Subhorizontal fiber-glass forepoles;

C: Twin steel ribs;

D: Steel fiber-reinforced shotcrete;

E: Lateral micropiles;

F: Subvertical jet columns;

G: Second work level;

H: Concrete invert;

I: Final concrete lining

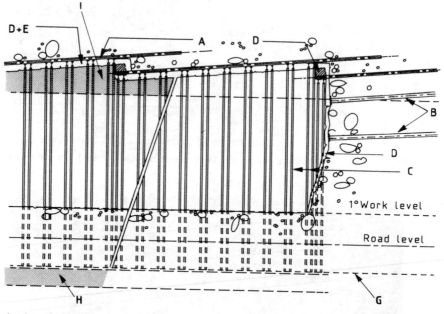

Fig. 4 Longitudinal section showing the advancement scheme using umbrella arch

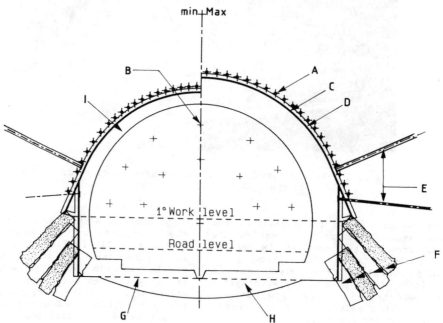

Fig. 5 Cross section showing the umbrella arch and other reinforcement

The umbrella pipes are 12m long. However, the excavated length is only 9m using 1-m steps, with the rib and shotcrete installed after each step. Thus, there is a superposition of 3m length between two consecutive umbrellas (Figure 4).

The excavation begins in the upper half section (first work level or heading). The structural support (steel ribs and shotcrete) is provided for immediate stability of the excavation. The next phase of excavation, benching, opens the lower half of the section (second work level). This creates the need for the following types of support (see Figure 5):

* Steel ribs (spaced at 1m) with wide footing and 30cm of steel-fiber reinforced shotcrete.
* Lateral micropiles which impede sinking of the floor and, above all, control the convergence of the benched section of the tunnel.

The benching phase should be based on consideration of the full section, including preliminary support of the heading, water proofing, and permanent lining of the entire section.

In order to reduce the stability problems around the lower section of the tunnel, the ground around the periphery of this section is preconsolidated using subvertical jet-grouted columns.

The presence of water, coupled with the presence of rock boulders, will cause instability of the face. A designed solution to this problem was to presupport the face with fiberglass forepoles. In addition, the unexcavated bottom part of the face, the "nose" (see Figure 6), provided additional confinement to the face.

4 CONSTRUCTION AND GROUND CONTROL

The construction scheme for the twin bores of the study tunnel involves the following steps (most of which are shown in Figure 7):
A. Installation of the umbrella arch in the heading using a two-arm machine
B. Installation of fiberglass forepoles in the face
C. Excavation by back hoe (in 1m steps and installation of steel ribs) for a total advance of 9m
D. Emplacement of fibre-reinforced shot-crete
E. Installation of lateral micropiles using a hydraulic drilling machine
F. Installation of subvertical jet columns
G. Excavation of the second work level (or bench)
H. Casting of the invert
I. Completion of the final, concrete lining

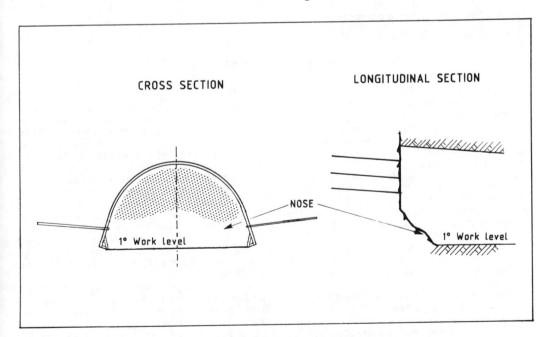

Fig. 6 Support of face with forepoling

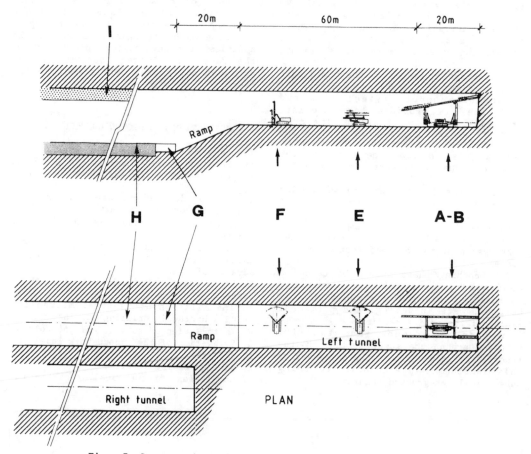

SECTION

Fig. 7 Construction scheme for the twin bores of the tunnel

We note that the advancing faces of the two bores are normally separated by 100m. This distance was dictated by the space requirements of the various operations. The effect of the location of the invert is discussed later.

The behavior of the ground was monitored with the progress of construction. This involved measuring deformation (of the two bores and the pillar) and loads on the spread footings of the ribs at the first working level.

Problems of excessive load on and deformation of the support occurred in both bores after an advance of about 250m.

A significant initial increase in pillar deformation is noticeable at two different intervals of face advance. First, in 30-40m of advance (about 1 month) of the face of the first bore, the deformation of the supports can be seen with the naked eye.

Let us say that the "impact distance" due to the first bore is 30-40m. Second, when the face of the second bore arrives within 20-30m of the face of the first bore. We note that the impact distance (and, in effect, the equivalent impact time) for the second bore is reduced to 20-30m because the second bore is being excavated in an already weakened material.

Likewise, the increase in deformation due to the arrival of the second bore is also larger because of the increased load on the pillar and the properties of the material having been reduced to the residual values.

A recommended remedy for the above problem is to keep the face of the second bore behind the permanent invert of the first bore (item H, Figure 7).

An example of visible deformation is shown in Figure 8. Two sets of deformation

52

Fig. 8   A detailed view of the excessive deformation of steel
ribs in station 32 (see Fig. 10)

measurements (in the pillar and the outer
wall) for the two bores are shown in
Figures 9(a) and 9(b). These figures
clearly demonstrate the significantly
higher deformation in the pillar compared
to that in the outside walls. The location
of the measurement stations is given in
Figure 10.

As part of any solution to this problem
one must think of strengthening the pillar
between the two bores as well as enlarging
it. The results of the recommended
enlargement of the pillar from the present
8m to about 30m are yet to be observed
(with the progress of the excavation).

In the transition stage (from 8-30m
thickness of the pillar) which involves
another 250m of excavation, the current
ground control procedures will continue.
These involve the following:

1. Stop the tunnel advance and install additional support in critical locations.

2. Install the temporary invert at the
first work level after each advance of 9m.
3. Install additional horizontal micropiles (Figure 11) to strengthen the pillar.
4. Increase the bearing capacity of the
ground at the foot of the rib by jet
grouting. Note that the 4 subvertical
columns on each side of a bore must not be
installed together, but staggered in time.
This will reduce the amount of vertical
deformation which is an inherent, immediate effect of the jet grouting operation.

5 CONCLUSION

The large deformations and loads encountered by the supports in the morainic
formations were anticipated in the design
of excavation of the two bores. However,
the excessive deformations in the pillar
(requiring additional support measures)

53

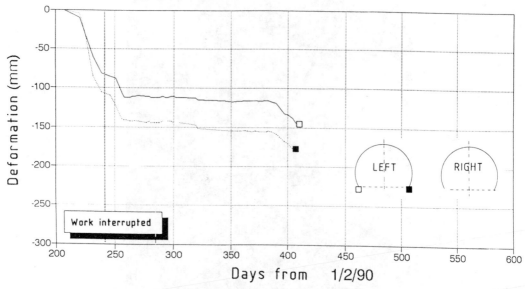

Fig. 9(a) Vertical displacement at the first working level on the pillar side and outer edge of the left bore

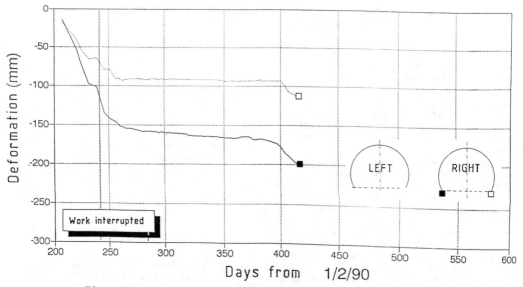

Fig. 9(b) Vertical displacement at the first working level on the pillar side and outer edge of the right bore

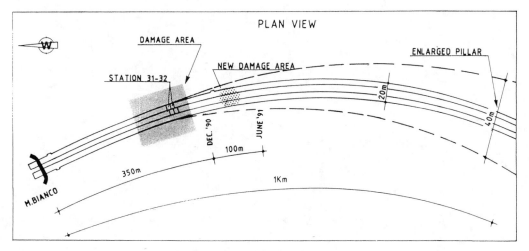

Fig. 10   Plan   view of the two bores of the tunnel showing the problem areas
          and   recommended enlargement of the pillar

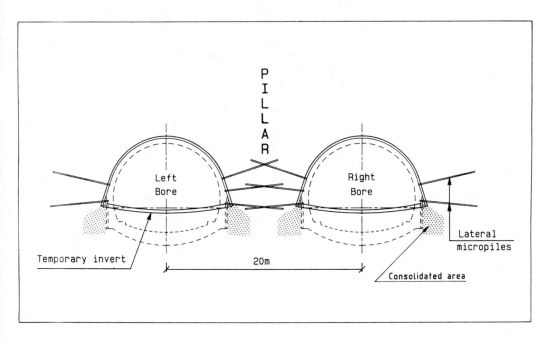

Fig. 11   Cross section of the two bores showing reinforcement of the pillar

resulted from the initially specified, narrow dimension of the pillar.
The recommended solution for minimizing the support requirement is twofold: (1) increase the strength of the pillar (widening and/or reinforcement) and (2) not advance the second bore beyond the invert of the first bore.

Effects of Geomechanics on Mine Design, Kidybiński & Dubiński (eds) © 1992 Balkema, Rotterdam. ISBN 90 5410 040 0

Soil strain measurements in areas of underground mining exploitation

J. Gustkiewicz, A. Kanciruk & L. Stanisławski
Polish Academy of Sciences, Strata Mechanics Research Institute, Cracow, Poland

ABSTRACT: The concept of a new method of measuring soil strains induced by underground mining exploitation is proposed. The method and an appropriate apparatus may be applied for long-term, extending for several years, measurements of infinitezimal and finite strains. Examples of results are given illustrating a local three-dimensional state of strain, expressed as a function of time and a function of a certain situation of an exploitation front with respect to a measurement site. The existence of some relations between a surface deformation and an underground tremor and between strain and the changes in the seismic wave velocity in soil and apparent resistivity has also been shown.

INTRODUCTION

At the Laboratory of Rock Strains of the Strata Mechanics Research Institute of the Polish Academy of Sciences in Cracow work has been carried out for over 30 years on the development of methods of long-term measurement of soil strains in areas of underground mining exploitation. Several variations of the measuring apparatus have been constructed which enable to determine flat, and even three-dimensional states of strains. After it has been installed in soil, the apparatus is able to work for a period of time as long as tens years. The apparatus may be also used to measure deformations of rock mass.

The movements of the surface of a territory are usually controlled by geodetic methods. The same methods, which are normally used to determine the quantities, are employed to define the changes in these quantities. e.g. deformations. The proposed apparatus, on the other hand, is intended to be used for measuring changes in the distance between some points, hence it is much more accurate and sensitive to this changes than an instrument for measuring the distance only.

Geodetic methods of measurements require relatively large measurement bases, i.e. distances between the neighbouring measurement points. The bases of instruments used for measuring the deformations may be in practice of an arbitrary length, thus, in a particular case, their size may be arbitrarily reduced, which is important in the case of significant strain gradients.

An apparatus the basic part of which is the strain gauge must be installed in soil for good. Thus it is often necessary to cover the observed territory with a sufficiently dense network of instruments, which is expensive and sometimes it may not be possible at all for some technical reasons. In such a case the geodetic methods enable to cover relatively large areas with continuous observation lines.

Electrical strain transducers may be buried in soil in practice at an arbitrary depth, in particular, below the freezing zone, 24 hour temperature fluctuations, as well as great changes in the soil humidity. This guaranties stable conditions for the functioning of the gauge whereas the measurement of a distance by means of a measuring tape or invar wire, which requires visual contact with the reference marks may be in great errors resulting from great fluctuations of temperature and humidity and soil freezing on the surface.

It may be said, that the application of strain gauges installed in soil is a substantial completion of geodetic observations. It should be also added that because of its sensitivity and accuracy the measurement mode using the strain gauges allows to notice changes in soil subjected to the effect of underground mining exploitation from day to day.

Thanks to instruments described further new results and experiences have been gained recently. The results and experiences are complementary to earlier publications [Gustkiewicz J., Klein G. (1977)

and Gustkiewicz et al. (1985)].

These data refer to the stability of the instruments during the many years period in which they are buried in soil, long-term measurements and extension of the measurement range. The latter is important for the case when the described apparatus is used to observe landslips. Illustrative examples obtained in the course of deformation measurements on the territories of mining exploitation are also given.

1 CHARACTERISTICS OF THE MEASURING APPARATUS.

When designing the strain gauges, attention was given to the fact, that they would be buried for many years in soil and would be exposed to disadvantageous effects of environmental factors. The possibility of installing the gauges within rock mass was also taken into consideration.

Two types of strain gauges have been constructed: TTCS 4000.3 and TTCS 2000. together with corresponding display systems among others, KA-3D. The description of the instruments and their accurate metrological characteristics were published by Gustkiewicz J. and Klein G. (1977) and by Gustkiewicz J. et al. (1985) respectively. The TTCS 4000.3 gauge may have its base in the interval 2-4m and the measurement range of 30mm, the TTCS 2000 -the base of 0.5-2m and the measurement range of 15mm. In case they are to be installed in soil both gauges are fixed to steel anchors with a surface of about 0.1m2. The pressure of the anchors on soil mass does not exceed 700N/m2. Fig.1 shows a general view of the TTCS 4000.3 gauge. In the case of the TTCS 4000.3 gauge the changes in the distance between the anchors are transferred through a string onto a double-arm lever, not seen in the picture, and in the case of the TTCS 2000 - on a system of levers. The movement of the lever or the system of levers produces a change in the tension of the vibrating steel wire of the gauge. An electromagnet which acts with the wire consists of a coil wound around a core and is fastened near the middle of the wire so that the distance between its pole shoes and the wire is about 1mm. The electromagnet plays the role of a system initiating the vibrations of the wire and also the role of a transducer of mechanical vibrations into electric oscillations. A short (about 2ms) rectangular current pulse is triggered on the electromagnet terminals. Then the wire generates expotentially decaying vibration of a frequency depending, among other things, on its tension. This vibration is transformed into an electric signal and

sent by cable to the display system. Each strain gauge is equipped with a temperature gauge, which consists of a massive beam with a wire stretched on it, and an electromagnet. The beam is made of a material whose coefficient of termal expansion considerably differs from that of steel. With changing ambient temperature this difference produces changes in the wire tension and consequently in the frequency of its vibration. Initiating the vibration of the wire and its transformation into an electric signal is realized by the electromagnet similarly as in the system of the strain gauge. The measurement of temperature is of great importance as the strain gauge is not free from the influence of temperature on its output signal. Such components of it as e.g. string change their length depending on temperature. However, before its embedding in soil each strain gauge is tested at various ambient temperatures in a special chamber with controlled temperature, and consequently, the measurement of temperature allows to correct the result of the measurement of strain. Moreover, a marked influence of temperature on the expansion of soil has been observed. It has been noticed, that the expansion of soil in the vertical direction is considerably greater than the expansion in horizontal direction. The results of these observations can be found in the papers by Gustkiewicz et al. (1985).

The vibrating wire meter KA-3D is a device which can be connected through a cable with the electromagnet terminals of the strain gauge or temperature gauge. It may also cooperate with other vibrating wire gauges. The measurement by means of this device consists in triggering a pulse exciting the vibration of the measuring wire. After a certain delay the duration time of the succesive 200 periods of the vibration is measured. After appropriate prescalling the result is shown on the display unit and it represents the duration time of a single period of the measuring vire vibration in microseconds.

One of the basic requirements which the instruments must satisfy is the long period of infallible functioning extending to tens of years. The investigation results of the instruments obtained so far seem to confirm this postulate. Gauges from a population of about 30 pieces were taken out from soil after 8 and 14 years since their installation. The gauges were subjected to renewed calibration by determining the dependence between the change in the distance between the anchors and the square of the frequency of the wire vibration. In nearly all the cases the dependences

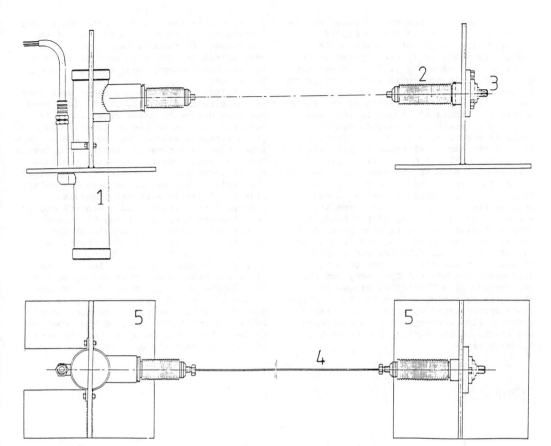

Fig. 1 General view of a TTCS 4000.3 gauge. 1. vibrating wire gauge, 2.device for range control, 3. handwheel, 4. string, 5. anchor

between these two quantities are practically coincided, which can be seen in Fig. 2. The maximal differences in the slopes of such strain lines as shown in the figure were 12%, were very infrequent and resulted from unnoticeable defects of the materials used for the construction of the instruments. It should be mentioned, that in the vicinity of the building of the Strata Mechanics Research Institute exists an experimental field on which several gauges have been buried in soil and the measurements of strains and temperature are taken in practice every day. These instruments were installed in December 1973, thus they have been burried in the ground for 18 years and there are no signs that they are not working correctly.

The measurement range of both gauges is 30 and 15mm respectively, which guarantees their high precision and sensitivity. To make possible the use of the instruments

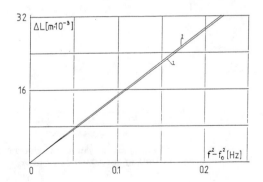

Fig. 2 Dependence between the vibration frequency of the wire and the change in the distance l between the anchors of a TTCS 4000,3 gauge.
1. before installation in the soil
2. after 14 years of staying in the soil

when soil deformations exceed these ranges, a string of controlled length has been employed. The length of the string is changed by means of a screw system in an air-tight enclosure to protect it against corrosion. This system is seen in the right hand part in Fig. 1. A turn of a handwheel causes a change of the length of the string within the limits of 120mm. In the case when, as a result of soil deformation, the gauge is approaching the range limit, a turn of the handwheel makes it possible to set up the device e.g. at the middle of the measurement range of the vibration frequency of the wire. To have acces to the handwheel it is necessary to make a special hole in the ground or to use a special wrench to enable turning of the handwheel from some distance. It is also possible to use an electric motor supplied from a car battery. Fig. 3 illustrates the functioning of such a construction in the form of dependence between the frequency and the deformation of a TTCS 4000. gauge. It shows the measuring points and the straight lines approximated by this points. Each straight line corresponds to a change in the length of the string by a whole measurement range.

former one is that it does not require any interference into the functioning of the apparatus, the disadvantage, however, is reduced sensitivity and taken as a whole - smaller measurement range than in the former case.

A digital vibrating wire meter described by Gustkiewicz J. et al. (1985) is a hand - operated device by means of which it is possible to measure the periods of the gauge wire vibrations. On the basis of this instrument an automatic registrating device CRPS has been constructed which realizes cyclic measurements without being serviced. The registrating device comprises a multiplexer, vibrating wire meter module, memory, output module and a control module. 16 vibrating wire gauges may be connected to the multiplexer. The module of the vibrating wire meter induces the vibration of the wire and measures its vibration period. The memory module allows to complete the measurement data arranging them into blocks which by means of the output module are recorded in an externally connected casette memory. The control module initiates the functioning of the remaining modules. The functioning of the registrating device consists of a cycling

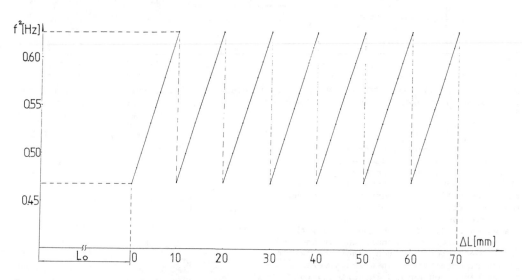

Fig. 3 The functioning of a gauge of the TTCS 4000.3 type equipped with a range control device including an electric motor

Another method of extending the gauge range is to include into the system a helicoidal spring in a series connection between the end of the string and the anchor. An advantage of this system in comparison with the

measurement of the vibration periods of the wires of all the gauges and recording the results on a casette. This process is repeated at some definite time intervals set up manually from 1 to 120 min.

2 INSTALLATION OF THE GAUGES IN SOIL

The gauges are installed in soil at a depth below the frost penetration in winter and below which the 24 hours temperature oscillation disappears. On the territory of Poland this value attains 1.2m. The gauges with horizontal measurement basis are usually installed at a depth of 1.5m. To install the gauges in soil it is necessary to make an excavations and next to cover the gauges with soil. In view of such disturbance of the original structure of soil it is necessary to wait for a period of time lasting several months till the changes resulting from soil subsidence and noticed by the apparatus are completed. Usually, the gauges installed so far allowed to measure deformations in the horizontal direction. As a rule they have been so embedded that they formed an equilateral rosettes comprising 3 or 4 measuring directions. This enabled to assign plane states of strains to the vicinity of the point on the soil surface. The following quantities are determined in particular: maximal and minimal linear strain, extremal shear strain and the directions of the main strains with respect to the northern direction. Fig.4 shows the excavation for an equilateral rosette comprising 3 measurement directions. In the ditches gauges can be seen while they are been covered with a mass of soil. 3 vertical pipes are visible which are used to insert a special wrench from the surface in order to change the length of the string between the anchors, as already mentioned. In the middle of the triangle formed by the ditches a cylindrical container is instal- led in which the cable terminals of the measuring system of the gauges are kept. The first attempts at determining a three – dimensional state of soil strain have already been done. In this case the flat horizontal rosette of the gauges was supplemented with a vertical gauge. The application of only one vertical gauge is in accordance with the adopted assumption that one of the main strains takes place in the vertical direction. Apart from such an assumption it is possible to calculate accurately the volumetric strain of the soil.

3 ILLUSTRATIVE EXAMPLES OF SOIL STRAIN MEASUREMENTS

In June 1988 several soil gauges were

Fig. 4 Excavations for equilateral rosette comprising 3 measurement directions

Fig. 5 TTCS 4000.3 gauge in a trench before
covering it with a mass of soil

installed in an area under which at a
depth of about 500m Hard Coal Mine "Piast"
is exploiting a coal seam, about 3m thick,
by a long-wall system. One of these rosettes enables to determine three-dimensional
states of strain as a function of time.
In this case it has been assumed, that one
of the main directions of the tensor of
infinitezimal strains coincides always with
the vertical direction. The two other main
directions and the corresponding main
strains are calculated on the basis of the
measurement in the plane horizontal
rosette comprising 3 gauges. The measurement results for an observation period of 2
years are presented in Fig. 6. The figure
shows the values of the main strains
as a function of time. The distances
between the points on the curves assigned
to the given date of measurement are the
measures of the shear strains, occuring
between the pairs of straight lines originally perpendicular and forming bisectrix
directions with the main directions. Thus
there are extreme shear strains with
respect to the direction. The values of

these strains as a function of time are
shown in Fig. 7. For the dates marked in
Fig. 6 and 7 with lines there have been
drawn in Fig. 9 in a vertical projection
several situations at the exploitation
front with respect to the location of the
rosette on the surface.
It is worthy of note, that earlier
tension in one of the principal directions
caused by the exploitation of the area in
1988 is changed into compression when in
the year 1989 the territory under the
rosette was already an exploited one. In
this region, as it is known, the convex
side of the subsidence trough is directed
downwards. The main strain directions
remain practically unchanged throughout the
entire period of observations. An exception
here is the initial period of a few months
during which under the influence of exploitation the soil began to move. It is to be
recalled, that the third main direction
is by assumption always vertical. As it is
known, the trace of the tensor of infinitezimal strains represents the volumertic
strain of the medium. The sum of the main
deformations as a function of time is
shown in Fig. 8, where sudden, almost jump-
like changes of strains with time can be
seen. At present it is not possible to
account for these changes in view of the
lack of an adequate amount of the measurement results of vertical deformations.
They may be connected with underground tremors which occur in abundance in the region
of the mine. They may also be due to temperature changes in the soil in a period when
these changes attain their greatest derivative as functions of time, e.g. in spring
and autumn. The interaction of these two
reasons is also possible.
The existence of relation between soil
deformations on the surface and underground
tremors is evidenced by the results published by Gustkiewicz et al. (1985). Starting
with the year 1974 deformations on the area
of Copper Mine "Lubin" in Legnica - Glogow
Copper District have been measured. In the
period 1976-1977 the movements of the
ground resulting from the exploitation
at a place where 5 rosettes had been
installed, were completed. Since August
1976 a uniform shrinking of the ground
recorded by all the rosettes has been
observed. On March 24th, 1977 there
followed an underground tremor at a depth
of 1250m and a magnitude of 4.5. Its epi-
centre was at a distance of 5 km from the
rosettes. The tremor was registered by all
seismic stations in the whole Europe.
Since the day of the tremor the ground has
been subjected to uniform tension reaching
the state which had been registered in July
1976. The details referring to the tremor

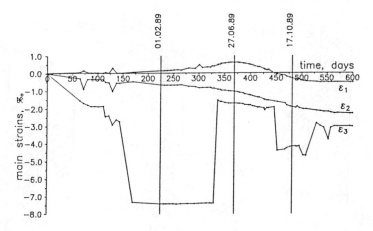

Fig. 6 Main, i.e. maximal, medial and minimal strains as functions of time.
Hard Coal Mine "Piast", Upper Silesia

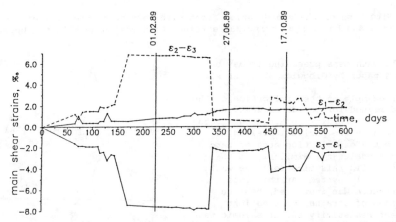

Fig. 7 Extreme shear strains as functions of time. Hard Coal Mine "Piast", Upper Silesia

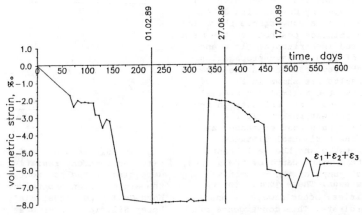

Fig. 8 Volumetric strain of soil versus time. Hard Coal Mine "Piast", Upper Silesia

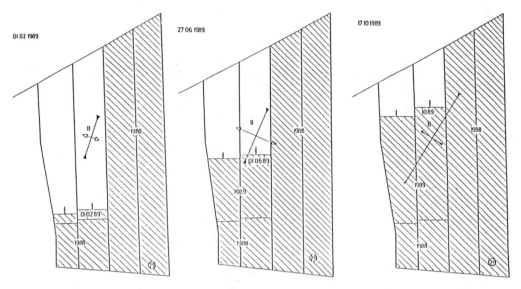

Fig. 9 Situation of the exploitation front with respect to the site of the rosette on the surface and horizontal principal strains. Hard Coal Mine "Piast", Upper Silesia

and to the phenomena preceding it may be found in a paper by Gibowicz S. J. et al. (1979).

Another example of the application of the presented method for the determination of strains are the results obtained during the measurements in the period 1983-1984 in soil above the exploitation area of Hard Coal Mine "Andaluzja". Similarly to the mine "Piast", in this mine coal was worked by the long-wall system. At the place where a rosette was installed, besides observations of strains, observations of apparent resistivity and of seismic wave velocity in the soil were carried out. Both these quantities were measured paralell and perpendicular to the direction of the wall. Complete results and their discussion can be found in a paper by Gustkiewicz J. et al. (1988). Here as a convergence there are given certain coincidences between the main strains measured in horizontal plane and the velocity of seismic wave propagation and the resistivity measured in the same plane. The results are shown in Fig. 10. The figure shows the maximal and minimal strains, the seismic wave velocity in the soil and the apparent resistivity. All these quantities have been expressed as functions of the horizontal distance between the rosette and the wall front. Worthy of note are the parts of the graphs where distinct jumps of the respective values can be seen. These jumps, for each of the quantities, occur always at one and the same coordinate. This convergence of changes is a confirmation of the correct-

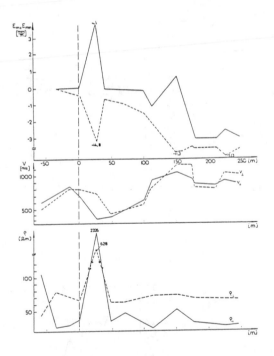

Fig. 10 Main strains, seismic wave velocity and apparent resistivity of soil as depending on the horizontal distance of the exploitation front from the place of Hard Coal Mine "Andaluzja", Upper Silesia - after Gustkiewicz et al. (1988)

64

ness of the response of each apparatus to the relative stimuli.

The above facts are evidence that a precis apparatus has been developed which may be used for long-term observations mainly of the effects of mining exploitation on the surface; it also creates a chance for the prediction of dynamic phenomena associated with mining exploitation.

REFERENCES

Gustkiewicz, J. & G. Klein 1977.
  On a Certain Strain Measurement of Soil
  Applying the Method of Strain Gauges with
  Vibrating Wire.
  Field Measurements in Rock Mechanics:
  123-135. Rotterdam: Balkema.
Gustkiewicz, J., A. Kanciruk
  & L. Stanislawski 1985.
  Some Advancements in Soil Strain
  measurement methods with Special
  Reference to Mining Subsidence.
  Mining Science and Technology 2: 237-252.
  Amsterdam: Elsevier.
Gustkiewicz, J., A. Idziak, L. Chodyn,
  A. Kanciruk, L. Stanislawski, A. Goszcz
  & W. Zuberek 1988.
  Zmiany predkosci fal sejsmicznych i
  elektrycznego oporu wlasciwego w wyniku
  deformacji gorotworu wywolanej
  eksploatacja gornicza.
  Publs. Inst. Geophys. Pol. Acad. Sc.,
  M-10 (213): 97-108.
Gibowicz, S. J., A. Bober, A. Cichowicz,
  Z. Droste, Z. Dychowicz, J. Hordejuk,
  M. Kazimierczyk & A. Kijko 1979.
  Source Study of the Lubin Tremor of
  24 March 1977
  Acta Geophys. Pol., 27(1): 3-38.

Effects of Geomechanics on Mine Design, Kidybiński & Dubiński (eds) © 1992 Balkema, Rotterdam. ISBN 90 5410 040 0

A research into underground pressure in longwall face induced by use of sublevel caving in thick seam

Li Hong-chang
China University of Mining & Technology, Xuzhou, Jiangsu, People's Republic of China

Zhu Shi-shun
ZhengZhou Coal Mining Machinery Factory, People's Republic of China

ABSTRACT: This report is written basically in accordance with the test results of two fully mechanized longwall sublevel caving faces. This paper mainly deals with the movement laws of roof coal and upper strata, and support loadings and types when coal face is under operation with a thickness of coal seam about 6 meters. It is believed that support loading in sublevel caving working faces do not surpass that in slicing coal faces which have similar roof and coal seam conditions. It is also proved that the caving supports with long canopy are best suitable for use in hard thick coal seams. If deep holes are drilled in the panel in parallel with faceline and are injected with high pressure water, the strength of coal body will be reduced, and the caved roof coal recovery will be increased.

1 INTRODUCTION

China has a large amount of gently inclined coal seams with a thickness from of 4.5-6.0 meters. The thick seams are widely distributed in many coalfields almost throughout the country. They are operated by longwall mining system. Three major mining methods to be used are as follow

1. Two slices operating system is adopted, wire mesh is used between those two slices.

2. High power supports, which working height ranges from 4. 0 to 5. 0 meters, are employed to mine whole coal seam once for all.

3. Caving power supports are adopted for mining the bottom slice, which thickness is from 2. 5 to 3. 0 meters , while the roof coal is recovered by sublevel caving.

The slicing mining system is very popular at present, because its cutting height of each slice is smaller, the technique to be used is relative simple, and the performances of mining induced pressure are not very intense. However, due to the long time between the operations of two slices, the artificial roof are very easy to be etched, and the cost of materials is high , and the ventilation and transportation systems are very complex. In recent years, high

supports have come into use for mining thick coal seams without any slice, a better effect of concentrative production has been obtained under suitable roof conditions. But, if there are local disturbed roof or soft coal seams, serious roof falls and coal wall collapses occur frequently, and the supports are very easy to be crooked and twisted. It is not a simple thing to adjust the crooked or twisted supports into a normal situation because of their big height and heavy weight, In actual operation , therefore , the working height of the supports has to be reduced, and the supports are working under an abnormal situation. The loss of coal reserves is thus increased.

Nowadays sublevel caving mining methods used in gently inclined coal seams are tested and are gradually popularized. This is a very important innovation of mining methods in thick coal seams. The initial test results indicate that if the conditions are suitable, the method has the advantages of higher productivity, higher efficiency, higher block ratio, less cost, less drain on materials, simpler system of conveying and ventilation, etc.

2 CONDITION FOR USING

Two major types of sublevel caving power supports are being used. One is shield type installed with one armored face conveyor. Because caved roof coal flows from the opening or shute door into the armored face conveyor, the opening is located in the shield beam at a high position, and the canopy is very short. This type of support is called high opening caving support. For example, typeZFD — 4000 sublevel caving support , which is employed by Wang Zhuang Mine of Lu An Mining Bureau, belongs to this kind of caving support . The other is chock shield type caving support equiped with twin armored face conveyors, Because the roof coal caves on the goaf side and flows into the rear conveyor through the door with a shutter, the caving door is lower in the shield beam, and the canopy is longer. thus the roof to be supported is longer. this type of caving support is called lower caving support. For example, type FD- 4400, which is employed by No. 1 Mine, Yang Quan Mining Bureau. The specifications of these type caving supports are shown in Table 1 and Figure 1.

The working conditions and coal seam natures of the two sublevel caving underground test coal faces are as follows: the former is face 4309 in 3# seam of Wang Zhung coal mine. The thickness of total coal seam, which contains 4 dirt bands with thickness ranging from 0. 02 to 0. 2 meter and with inclination 5-7 degrees, is 7. 26 meters. the strength of coal body ranges from 10-25 Mpa. The working face is located at a depth of 145 meters below the earth surface, and the face length direction is 120 meters. The cutting height of bottom slice operated by shearer with supports ranges from 2. 8 to 3. 0 meters. Figure 2a shows the natures of roof and floor.

Tabel 1   Sublevel Caving Support Specifications

Model	ZFD4000-17/33	FD4400-16. 5/20
Height range(m)	1. 7-3. 3	1. 65-2. 60
Max. operation height (m)	3. 0	2. 6
Mini. operation height (m)	2. 60	1. 65
Width range (m)	1. 43-1. 58	1. 43-1. 60
Yield load (KN)	4000(32Mpa)	4400
Setting load(KN)	3600(28Mpa)	4000
Canopy support density(KN/m·m)	763	518
Opening dimension in plan (mm)	2030·820	1500·900
Length of canopy (m)	1. 37	3. 29
Caving shute swing upwards	0°	45-84°
downwards	45°	18-47°

The later where test was carried out is working face 8605 in the No. 15 coal seam, No. 1 Mine, Yangqing Mining Bureau. The average thick-ness of seam is 6. 05 meters, the dip ranges from 2 to 11 degrees, the strength of seam ranges from 20 to 25 Mpa, the width

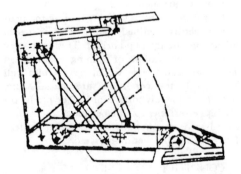

Fig. 1 a.   ZFD4000-17/33 shield support

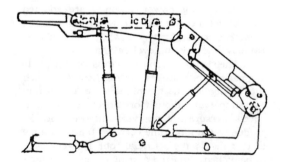

Fig. 1 b.   FD4400- 16. 5/ 26 chock shield support

Nature	section	thickness
sandstone		12.0(m)
clay with sandy shale		3.03
3# coal seam with 4 dirt		7.26
sandy clay		2.87

Nature	section	thickness
limestone		10.62(m)
black shale		1.72
15# coal seam with stone		6.05
white sandy stone		2.8

a. No. 3 coal seam, Wangzhuang
b. No. 15 coal seam, No. 1 Mine, Yangqing

Fig. 2   Columnar section of coal seam and strata

of coalface is 42 meters, the working depth is from 150 to 350 meters, and characteristics of roof and floor are shown in Fig. 2b.
Two main purposes to be reached in controlling the surrounding strata or rocks of sublevel caving coal face are:

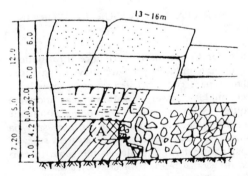

Fig. 3a The movements of roof coal and overlying strata

1. Suitable breaking of roof coal should be ensured within the range of controlled roof, and the caved roof coal blocks can flow out through the opening or shutter door without a hitch. It is also ensured that the collapse of overlying strata should keep step with the caved coal falling down without forming a big hole. Thus, the roof impacting on supports will be reduced to the least pressure.

2. A better integrity of unsupported roof coal between canopy tip and faceline should be kept, so coal production will not be broken or interposed by serious tip roof collapsing or falling.

3   THE DISTINGUSHING FEATURES OF THE BREAKING AND COLLAPSING OF ROOF COAL AND OVERLYING STRATUM

There was a used road on strike in the top slice 15 meters from the tail road in panel 4309. After repaired, it was used for measuring the mining induced pressure and observing the movements of roof coal and overlying strata. The crevice drawings and their parameters from ribside of roof coal and overburden strata were taken at different distances from coal wall before faceline. Figure 3 shows the measured results.

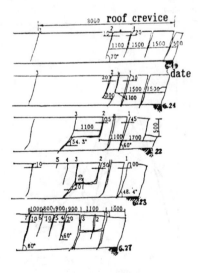

Fig. 3b   The relations between the breaking of roof coal and the advance of coal face (Enlarged A Area)

As shown in Figure 3, the roof coal begins to split open at 8 meters before the faceline, then it is inclined towards working face.

The crevices are with gradient of 60 – 80 degrees from horizontal inclined towards goaf. In accordance with the advancing of coal face, the crevices extend downwards and their inclinations decrease. Meantime, the density of crevices is increased. The distance between crevices ranges from 0. 8 to 1. 5 meters.

In order to determine how the forward abutment pressure acts on the roof coal and the overburden strata, two indirect measure methods are adopted.

1. Three observing cross sections with 13 holes were set up in the tail road. the depth of each hole, with a diameter of 42 mm, was 2. 5 meters. A supersonic instrument was used to detect the surrounding strata crevices by measuring the supersonic wave propagation ratio in the holes at different positons before the faceline. The measured results are shown in Table 2.

Table 2   The measured results

Before faceline(m)	30	25	18	13	12	8	5	4	2
Average wave speed (m/sec.)	1097	1105			1098	1127	1093		911
Loading density of supports (Mpa)			8. 4	10. 2		16. 8			12. 3

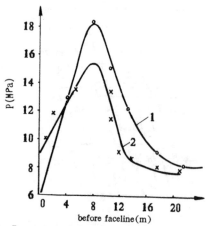

1. In main road   2. In tail road

Fig. 4   The curve of working resistance of support in **main and tail roads**

2. Four observing stations were set in the main road and tail road Hydraulic pressure automatic recording instruments, which were connected to the single hydraulic props, were used to measure the relations of support loading density and the advanced coal face. Fig. 4 and Table 2 show the underground test results.

The above mentioned results have shown no difference from the development laws of roof coal crevices. These indicate:

1. Along with the advancing of working face, the forward abutment pressures increase gradually and will reach the Max. value at the head of 10 meters before faceline in coal seam, and the overlying roof strata break initially at the head of 8 meters before the faceline.

2. At the head of about 4m in seam before faceline, immediate roof breaks and then **splits open fully**, generating 2-3 descending roof steps with little staggering height. The width of crevices ranges from 5 to 20 mm. The broken or cracked immediate roof strata collapses owing to its weight itself.

3. The breaking and tuning of main roof cause roof coal to generate crevices with interval distances of 10 — 100 cm, the width of crack is from 5 to 20 mm. The closer the crevices approch the faceline, the more quickly the crevices develop. The horizontal displacements of broken roof coal are about 200 ~400mm.

4. Roof coal breaking is mainly done by the interaction of upper roof strata and roof coal. Frequent movements of supports and their structures disturb the ballance of broken roof coal and break big blocks into small ones , thus the recovery is being increased.

4   STRATA BEHAVIOUR IN WORKING FACE

In order to get the performance information of mining induced pressure in working face, three observing stations were set up at the upper, middle, lower   working face, and were equip   with pressure — descending automatic recording devices to measure the necessary parameters. Because the roof natures, support load parameters, are almost the same in two testing working faces, the features, of which the overburden hard strata break and generate pressure periodically, is very important for determing the structural and mechanical parame-

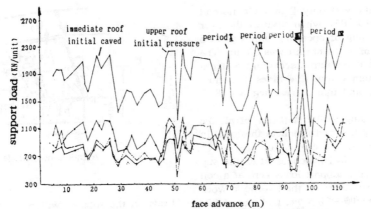

Fig. 5 Support working load changes in panel 4309 Wangzhuang Coal Mine
——total load — · — front leg load — · · —rear leg load —×— roof pressure

ters of support. Fig. 5 shows the changes of support loads in panel 4309. The results indicate:

1. According to the statistics, the thickness and strength of upper strata in panel 8605 are les but the length of canopy is longer than that in panel 4309. During the periodical pressure, the support resistances are 36% and 15% over the average resistance respectively. Table 3 shows the resistance statistical results of two sets.

Table 3 The resistence of two sets of supports

Parameters	initial pressure		periodical pressure		load in periodical(kN/unit)		dynamical pressure ratio
	pace (m)	average load (kN/unit)	pace (m)	frequency (times)	average	max.	
Panel4309	52	2402	14.1	4	2036	2750	1.15
Panel8605	19.2	2509	14.3	8	1831	3373	1.36

2. The support setting load Po, the time weighted average load Pt, and the cycle end load Pm, are shown in Figure 6. and Table 4. Each average value Po, Pt, Pm of first testing face is 1546 kN, 1866kN and 2264 kN, If the results calculated by $Pt+2\sigma_t$ or $Pm+\sigma_m$ are regarded as the consult yield loads of the support, the value is from 3240 kN to 3670 kN respectively. Actually, the designed yield loads are higher than needed.

3. The yield valves did not open and the descendings of legs were very little. The leg

Table 4 The resistence statistical results

Testing faceNº		resistance (kN/unit)	Standard deviation (kN)	Deviation ratio (%)	To yield load ratio (%)
I	Po	1546	887	57	39
	Pt	1866	900	48	47
	Pm	2264	975	43	57
II	Po	1469	443	30	33
	Pt	1636	481	29	37
	Pm	1793	584	33	41

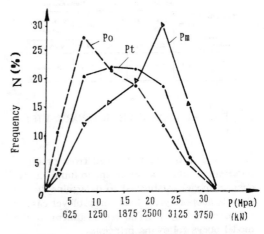

Fig 6 Frequency distribution of support loads

71

cycle descending was only 2 mm (in normal time) in panel 8605. when caving, the average leg descending of operating unit was 3.5 mm, and that of adjacent unit was 2 mm. The average of the maximum descendings of front leg was 5.6 mm (from five times of periodical pressure), and that of rear leg was 8.7 mm.

4. According to the statistics, of all the support load cycle—performances the loads increase one time in a cycle and occupy 79.1%. This shows that 10-15 seconds after the supports were set, the support loads kept at a constant, in which 54.7% of the hydraulic working pressure are 10—20 Mpa, and that of 20-36 Mpa occupies 10.7%, The support loads increase two times in a cycle, possess 7.2%, and it was caused by the loosing of roof coal and the suspending of immediate roof strata.

5. Dynamical pressures, which act on caving supports in sublevel caving coal faces, are not very high. Although the total mining thickness is two times the cutting height of 2.5-3.0 meters in bottom slice, the support loads in sublevel and slicing coal faces are almost the same. Comparing panel 8605 with other sublevel slicing coal faces which have similar roof and coal seam condition in the same seam, as shown in Table 5, the parameters of pressure and pace have bigger differences only in the period of initial roof pressure.

Table 5　Comparing the parameters of pressure

Time period	periodical pressure			initial pressure	
	pace(m)	Pma.-Pmin. (kN/unit)	dynamic ratio	pace(m)	Pmax.-Pmin. (kN/unit)
sublevel	14.3	3373-2552	1.36	19.2	3540-1372
slice faces (bottom)	9.7	3251-1539	1.34	26.0	2833-935

5　THE REASONABLE SUPPORT YIELD LOAD

According to the data obtained from practical observation and measurement in mine, a mechanics model established to determine the reasonable support yield load for sublevel caving is shown in Fig. 7 . The establishment of the model obeys following principles:
1. Break-lines of roof coal and strata are generated in front of the coal face, and the

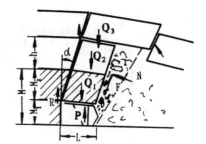

Fig. 7　A mechanics model

loads on the support will reach the maximum when the break-line arrives at the face line. This is the caculation basis for determining reasonable support yield load.

2. The break-face of roof coal in the goaf goes up with an inclined angle backwards and is supported by broken rocks in caved-goaf, and there is a friction resistance on the face between the roof coal and broken rocks.

3. The load Q_3, which is the main force to break roof coal and strata and enforce them to collapse, is generated when the overburden ballance structure formed as beam blocks with descending steps loses its stability. Although the sublevel caving has a bigger caved height and there are some slidings between blocks caused by shear stress, the beam blocks in normal caving strata band can also form a stable ballance structure by press force between blocks . So the influence load from disturbed roof strata which is under stability is about $1.3 \times 1.4(Q_1 + Q_2)$, where 1.4 is dynamic loading ratio and 1.3 is dispersion ratio.

4. The utilization ratio of support resistance is 75% , that is, the safety ratio is 1.3. Because the sublevel caving is usually used in coal seams with roof under stability, the moment of force generated by outer load will be ballanced by the internal distributed force of support structure.

5. The resistance is based on the maximun support length before caving. In fact, practical observation and measurement indicate that the load on support will be reduced when roof coal caving is operated.

Based on above, following equation is obtained:

$$P = K(Q_1 + Q_2 + Q_3 - R - N * Sin\alpha - F * cos\alpha)/L \qquad (1)$$

where p is reasonable support density, kN/m^2; K is utilization ratio of support resistance, K = 1. 3; L is supported roof coal length; N is press force from caved rock in goaf; F is friction between roof coal and caved rocks; F = N * tgφ; h is caved height of immediate roof, about 2M; Q_1 is weight of roof coal, $Q_1 = M_2 * L * \gamma_m$; γ_m is the unit weight of coal, about $14kN/m^3$; Q_2 is weight of immediate roof, $Q_2 = h * L * \gamma_d$; γ_d is the unit weight of rock, about $25kN/m^3$; Q_3 is added load when ballance structure has lost its stability;

Considering that support is under the maximun support length, then R = 0, the press force from the goaf rocks is:

$$N * \cos\alpha = \gamma_d * (M_2 + h)^2 *$$
$$* (1 - \sin\varphi)/2(1 + \sin\varphi) \qquad (2)$$

$$P = - 10. 5M^2 + 169. 5M - 91. 1, \qquad (3)$$

when $M_1 = 2. 8m, M = 4 - 10m, \alpha = 20°, \varphi = 35°; P = 418. 9 - 553. 9kN/m^2.$

6 SOFTENING ROOF COAL WITH WATER INJECTION

In order to increase the breaking rate of roof coal before caved, forward water injection was employed in Panel 4309. The holes were drilled 1. 5 meters apart from the immediate roof in ribside of the used tail road at top slice . The holes were designed to interset two main groups of crevices in the seam and to promote a softening effect of coal body. The holes were with 70 to 75 degrees from the road axis towards goaf. Their length were 70 to 75 meters (about 2/3 of the length of coal face). Thus, the injecting water could be avoided to leak to the main road. The interval space between holes were 15 to 20 meters, and the diameter of each hole was 42 mm. The holes were sealed with cement mortar, and the sealing length was 25 meters.

In order to determine the mutual relations of water injection effects with the pressures of water and surrounding strata, some experiments have been done in laboratory. the results show that higher the confinement pres-

sures are , the lower the water-to-coal ratio is. However, the higher the water injection pressures are, the higher the water-to-coal ratio is. The relations can be formularized as follows:

$$T = 2. 6156\exp(- 0. 517\sigma_z + 0. 714P) \qquad (4)$$

Where T is water — to — coal ratio , milli — Darcy ; σ_z is confinement pressure, Mpa ; P is water injection pressure, Mpa.

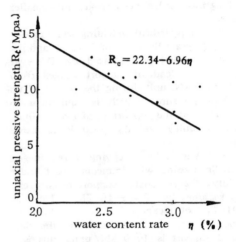

Fig. 8 Relations between water content strength of 3# seam

Figure 8 shows the relations of water content rate and uniaxial pressive strength in 3# coal seam is 3% after water injection the maximum can reach 5. 5%) , 1% — 1. 5% of water content rate are over the original moisture in coal seam. But the water injection should be kept for more than 15 days so that the coal seam can be wetted through. The flow rate of water injection pump is 2 cubic meters, and the water injection pressure in practice is 2 — 3. 5 Mpa. Average water injections are 200 tons/each hole, and the actual water content rate is 2. 71% . The coal uniaxial pressive strength is reduced from 17. 2 Mpa to 10. 5 Mpa, it is only 61% of the original strength. The coal recovery rate is increased to 79%. Morever, previous water injection can reduce coal dust consistency when caving , and prevent the combustion of lost coal in goaf.

7 CONCLUSIONS

Based on the foregoing, following conclusions can be reached:

1. The initial breaking and turning of roof strata occur at 8 m before front of faceline and enforce roof coal to generate inclined crevices with interval distance of 0. 8-1. 5 m. Along with the advancing of coal face, the crevices extend gradually and become larger .The amount of horizontal displacement of roof coal is more than that of vertical displacement. When the roof coal gets to the top above the working face, it has been broken into smaller blocks.

2. The periodical weighting width of two working faces are the same of 14. 2m, and the weighting strength is 1. 15 to 1. 36. The average cyclend load of support ranges from 2750-3373 kN/unit during the period of upper roof pressure, which is equivalent to 69%-77% of the support yield load (4000-4400 kN/ unit). So the yield loads are enough.

When the sublevel caving is used in gently inclined seams with immediate roof under stability, the reasonable support density can be determined by: $P = -10. 5M^2 + 169. 5M -91. 1$, coal thickness is from $4-10$ m, the support density is $420-560$ kN/m^2, the yield load of support is 4000 kN/ unit, this is enough for sublevel caving.

3. Comparing the sublevel caving coal faces with the slicing coal faces which are with similar roof and working conditions, although the average periodical weighting width in sublevel caving panel rises from 9. 7 to 14. 3 m (47% of it are increased), the support loads are almost the same.

4. The shield support with short canopy, which is equipped with one face conveyor, has higher support density at tip face, and the accidents from face conveyor are less. It is suitable for use in softer thick coal seams. The chock shield support with longer canopy , which is installed with twin armored face conveyor, has a larger working space and coal loading capacity. But the roof above at gate ends is more difficult to be controlled. It is suitable for use in harder thick coal seams.

5. Forward long—hole water injection parallel with faceline to soft coal seams is advantageous to break roof coal , to increase roof coal recovery, and to reduce the consistency of coal dust. In panel 4309, the water content ratio in coal seam was increased by 1%-1. 5% , and the strength of coal body was reduced by 39%, the roof coal recovery rate was also raised.

6. Each annual output of the two caving coal faces would be over 1 million tons. Comparing the sublevel caving coal faces with slicing ones, the output and productivity will be increased, tunnelling rate will be reduced, the cost of material and manual force will be saved , and the operation is more safe. So the underground tests are very successful, and this mining method should be popularized.

REFERENCES

Li Hong — chang, 1986. The correlation between ascending mining and stability of the overlying strata. The AusIMM 111awarra Branch, Ground Movement and Control related to Coal Mining Symposium: 232 — 239.

Li Hong- chang, 1985, Interaction between support and surrounding rocks in multi- slice longwall mining. Mining Science and Technology China coal Industry publishing House Transtech Publication: 39 — 46.

S. S. Peng Li Hong- chang, 1987. Large Scale Simulation Model as Longwall Design Tool. A Case of Shallow Longwalling. Longwall USA Conference papers: 267 — 296.

S. S. Peng, J. Wu, Li Hong—chang, 1986. How to determine yield load of longwall roof supports. Coal Mining, October 40 — 43.

Chien, M. G. 1982. A study of the behaviour of overlying strata in longwall mining and its application to strata control. Symposium on Strata Mechanies, University of Newcastle on Tyne: 13-17

Z. M. Jing, 1991. Application of sublevel caving with discharging opening hydraulic powered support to medium thickness hard Seam. Coal Science and Technology, 2 (n Chinese).

2 Theoretical and practical studies

A theoretical basis for physical models

N. Brook
Department of Mining and Mineral Engineering, University of Leeds, UK

ABSTRACT: A method of designing small scale models of mine openings is developed. Different scale factors can be used for rock strength, steel supports, yielding elements and rock bolts. Each scale factor reflects the most significant aspect of behaviour and this allows freedom of choice of materials. Methods of measuring the strength and performance properties involved are described. The application of the method to specific site conditions is considered.

INTRODUCTION

The use of small scale physical models is widespread in many branches of engineering, but perhaps the most effective use of models is in aspects of hydraulics. A broad consensus of opinion on the theoretical basis exists, depending largely on the use of dimensionless groups of the variables involved. The most frequently used of these is the "Reynolds number", which considers the ratio of inertia forces to viscous forces, and making this equal in the model and the full scale or field version of the equipment generally ensures comparable patterns of flow. For other effects other groups of variables are used, such as the ratio of inertia to gravity forces, which gives the "Froude number", useful in the study of the settling of solids in a fluid. Well over fifty such numbers are defined, and each seeks to model a selected physical effect. The continued use of dimensionless groups where it is not actually required is perhaps an occupational disease of hydraulic engineers, and can be counter - productive as indicated by Brook, 1987. When a physical analysis of the system is not evident, the dimensionless groups can be assembled using the physical dimensions of the variables as a guide, using such methods as "Buckingham's Pi theorem". This is, however, no substitute for physical analysis of the situation and can lead to meaningless groups of variables. The situation is further complicated in rock mechanics where many of the variables have the dimensions of stress. It is not possible for field and model systems to always have the same ratio for size, tensile strength, compressive strength, modulus of elasticity, density, strength of support members, and so on. The section below develops five similarity numbers which are based on making the field and model systems have the same physical effect for a selected aspect of the performance. The use of a chosen size scale does not imply a fixed scale for strength of the rock to model material, as in systems designed by the Buckingham Pi method by Hobbs 1966, and a variety of combinations of the properties of supports equipment can be considered.

1. Similarity numbers for physical models.

1.1 Size

The size scale for the model is chosen to suit the testing equipment available. Small size models are often easier to construct, but with larger sizes it is easier to model the structural features

of a rock mass. Bedding planes in
particular, for coal mine strata, are
extremely important in controlling the
behaviour of a model under load. The
failure to model significant structural
features can make the model completely
unrealistic. Not every feature can be
conveniently modelled, and structural
details such as cleat and joint gouge are
difficult to include. The size similarity
number, or scale factor, is simply the
ratio of the size of the field site to the
model, using any convenient dimensional
length. For such a typical length L,
model length L_m, the scale is thus

$$N = L/L_m$$

1.2 Rock strength

Many aspects of rock strength could
be considered, but most failure of
underground openings is due to the
removal of support of the rock, allowing
expansion into the excavation. In the
early stages of failure, the critical
behaviour is the tensile strain due to
compressive forces. For a selected part
of a site including an excavation the
loading will be force F, giving a
compressive stress of F/L^2. For a
material with Poisson's Ratio ν, Young's
modulus E, the tensile strain is then

$$\epsilon_t = \frac{\nu F}{E L^2}$$

This can be compared to the critical
tensile strain at failure in a tension
test, ϵ_{tc}, where To is the tensile
strength and

$$\epsilon_{tc} = \frac{To}{E}$$

The rock strength similarity number
can then be defined as N_R, where

$$N_R = \frac{\epsilon_t}{\epsilon_{tc}} = \frac{\nu F}{E L^2} \frac{E}{To}$$

$$N_R = \frac{\nu F}{L^2 To}$$

The value of F for the field site
can be found from the ground stress
and the area to be modelled, and the
model force F_m is controlled by the
loading system. For N_R the same
for both field and model then

$$F_m = F\left(\frac{L_m}{L}\right)^2 \left(\frac{\nu}{\nu_m}\right)\left(\frac{To_m}{To}\right)$$

From this it can be seen that using
a stronger model material increases
(To_m/To) and simply results in a
larger model force being required to
cause failure. Putting $L_m/L = 1/N$
gives

$$F_m = F\left(\frac{1}{N^2}\right)\left(\frac{\nu}{\nu_m}\right)\left(\frac{To_m}{To}\right)$$

1.3 Steel supports

Arches, bars and straps used to
support excavations are frequently
deformed in service, and the critical
behaviour can be considered to be
yielding of the steel due to bending.
The bending moment applied to
supports will depend on F x L, and
for a section modulus Z the
maximum stress due to bending will
be

$$S_b = \frac{FL}{Z}$$

This can be compared to the yield
stress of the steel, Y, to give a steel
support similarity number Ns, where

$$N_s = \frac{S_b}{Y}$$

The necessary model support section
modulus Z_m, is then

$$Z_m = Z\left(\frac{Y}{Y_m}\right)\left(\frac{L_m}{L}\right)\left(\frac{F_m}{F}\right)$$

putting $L_m/L = 1/N$, and using the
ratio of forces determined by N_R above,

$$Z_m = Z\left(\frac{1}{N^3}\right)\left(\frac{Y}{Y_m}\right)\left(\frac{\nu}{\nu_m}\right)\left(\frac{To_m}{To}\right)$$

By selecting a suitable value of Y_m in
relation to Y, the section modulus can
be designed to a suitable value. It may
be necessary to try and obtain the same
ratio of main to transverse section
modulii of the field and model supports,
which is readily achieved by controlling
the geometry of the model section. Too
much variation in the dimensions of the
support compared to the excavation

should be avoided, as this would affect such aspects as the area in contact with the rock.

1.4 Rock bolts

The performance of rock bolts can be defined in a number of ways, but although tensioning of the bolts is possible in some designs, giving active support, the main effect is passive support developed by the movement of the rock mass. The most used measure of the effectiveness of a rock bolt is the "pull out" force, but the bolt stiffness, S, determines the force developed. The bolt force F_b is thus

$$F_b = \epsilon_t \ L \ \times \ S \ = \ \frac{\nu \ F \ S}{E \ L}$$

This can be compared to the force applied to the bolt as ground stress, which is controlled by F. The ratio F to F_B is the rock bolt stiffness similarity number N_B and

$$N_B \ = \ F/F_B \ = \ \frac{EL}{\nu S}$$

The required stiffness of a model bolt is then

$$S_m \ = \ S \ \left(\frac{\nu}{\nu_m} \right)\left(\frac{E_m}{E} \right)\left(\frac{L_m}{L} \right)$$

$$\text{or } S_m \ = \ S \left(\frac{1}{N} \right)\left(\frac{\nu}{\nu_m} \right)\left(\frac{E_m}{E} \right)$$

The Young's modulus used here is that for the rock or model material, and not that for the bolt. If it considered that the pull-out force is the important aspect, defined as F_u, this can be compared with the applied force F, to give the model pull-out force as

$$F_{um} \ = \ F_u \ \left(\frac{F_m}{F} \right)$$

$$= \ F_u \left(\frac{1}{N^2} \right)\left(\frac{\nu}{\nu_m} \right)\left(\frac{To_m}{To} \right)$$

It may be possible to model both bolt stiffness and pull-out force with suitable materials.

1.5 Yielding supports

The use of yielding elements in support systems avoids severe damage due to imposed rock stresses, and by allowing a broken zone of rock to form near the excavation actually reduces the forces which are applied to the support member.

Typical items are yielding "stilts" for near vertical parts of roadway supports, and yielding hydraulic roof support chocks or shields on longwall faces. The ratio of the force applied to the member to the yield force can define a yielding support similarity number N_y, where

$$N_y \ = \ \frac{F}{F_y}$$

The design yield force for a support member in a model is then

$$F_{Ym} \ = \ F_Y \left(\frac{F_m}{F} \right)$$

$$= \ F_Y \left(\frac{1}{N^2} \right)\left(\frac{\nu}{\nu_m} \right)\left(\frac{To_m}{To} \right)$$

2. Determination of physical quantities involved

2.1 Load due to strata pressure

The value of the total force, F, which is applied to the site under investigation cannot be measured directly. A usual method is to calculate the stress due to "cover load" of the overlying rock to find the vertical component of load. The horizontal components are more difficult to assess, as sometimes the horizontal stress exceeds the vertical stress, this being true particularly for shallow depths of working. Ideally the in-situ stresses should be measured by, say, inclusion stress meters, but as these are expensive and technically difficult to use, the assumption that the horizontal stress equals the cover load stress is reasonable for mines in new areas, but for situations in previously worked areas the horizontal stress may be low, and due only to the effects of confinement of the rock mass. During mining operations the load applied to a site may increase greatly due to the removal of supporting material, this effect being mainly gravity induced and so the horizontal stress is not affected as much as the vertical stress. For model systems more than one loading arrangement can be studied, and it is prudent to finally achieve the maximum loading conditions envisaged at a site, this perhaps including heavy horizontal stresses as compared to the vertical stress.

2.2 Tensile strength of rock and model material.

The direct testing of rock for tensile strength is not always possible, especially if only broken fragments of rock are available from coring or hand sampling. Indirect methods offer a more convenient way of finding the strength, but since the correlation between different methods is not perfect the type of test chosen should be applicable not only to the rock involved but also to the model material, which is often a sand/plaster mixture very much weaker than rock. Two indirect tests are particulary suitable, the Point Load test and the tensile strength by bending, or modulus of rupture test.

The point load test can be applied to a wide variety of shapes and sizes of specimen as shown by Brook, 1985, and the results standardised to the usual Is(50) value by shape and size correction factors. Since the size correction factor depends to some extent on the material being tested the size correction should be kept to a small amount by using test specimens all of near the same size for both the rock fragments and test pieces of the model material. The force required to cause failure of the specimen for weak model material can be very low, down to about 50 N in some cases, and the test rig used should be able to measure such low forces accurately. The normally available apparatus which uses hydraulic pressure to apply, and to measure, the force is not sufficiently sensitive for this low strength type of test. The main features of the test are illustrated in figure 1.

The conventional modulus of rupture test requires a "beam" of the test material to be prepared. This can be convenient for materials which can be cast to shape and size, such as concrete or the sand/plaster type of model material often used,

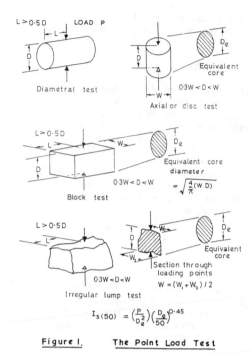

$$I_{s(50)} = \left(\frac{P}{D_e^2}\right)\left(\frac{D_e}{50}\right)^{0.45}$$

Figure I.   The Point Load Test

but standard rectangular section beams are difficult to prepare from rock if this is fragmented. A convenient alternative is the "disc test", a hybrid version of tensile strength by bending and the point load test, as described by the author (1982 and 1990). Support of the "beam" is by two steel balls and loading is by a third centrally placed ball, as indicated in figure 2. The use of ball points for loading means that the specimens do not need to be machined smooth so that natural rock parting surfaces are adequate. The disc can be treated as a rectangular section beam and T_{mr}, the modulus of rupture calculated from the usual formula.

$$T_{mr} = \frac{3}{2} \frac{PA}{bd^2}$$ with P the load, A the span b the breadth and d the depth.

As with the point load test, testing of model material requires a sensitive testing machine, as the failure loads may be as low as 20 N. To avoid size effects and other geometrical effects the test specimen geometry for rocks and model material should be similar, although the disc test does not have a prominent size effect. The test does

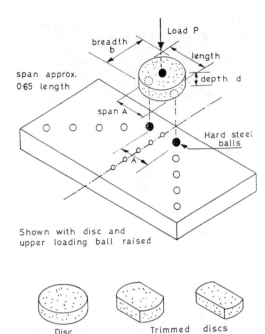

Shown with disc and
upper loading ball raised

Figure 2. **The Disc Test**

not give the true tensile strength, due
to the assumption used in deriving the
formula not being true for rocks. If the
different modulus of elasticity of rocks
for tension and compression is
considered, a value close to the direct
pull value can be calculated, as
described by Brook (1990).

2.3 Elastic constants.

The measurement of Young's modulus,
E, and Poisson's Ratio, ν, is generally
done by either fitting strain gauges, or
transducers, to a compression specimen.
A large specimen, of 50 mm diameter, is
generally required and sometimes this
may be difficult to procure. Poisson's
Ratio varies with the applied stress if
porous rocks are tested, and a
generalised value may have to be used
in the equations. Hanif (1974) found
very little difference in the values for
coal measures rocks and the model
material used, the value for both types
of material being in the order of 0.2.

2.4 Roof support strength.

The quantities required for designing
the model roof support members are
readily measured in most material
testing laboratories, and values such

as yield strength of steel can often
be obtained from suppliers. With a
very wide selection of materials
possible for model supports, such as
brass, lead, copper wire or wood,
little difficulty has been encountered
is finding suitable equipment.

3. Comparison of field and model
results.

Comparison of performance
measurements have been made at
sites in coal mines in order to
validate the model systems, the most
comprehensive being by Rayan (1980)
and Brook (1982). This series of
tests involved the construction of
layer for layer models for a site at
Fryston Colliery, by kind cooperation
of the (then) National Coal Board,
which were tested by steadily
increasing the vertical load whilst
rigidly confining the model in the
horizontal directions. The field
investigations involved roadway
deformation measurements and the
measurement of in-situ and mining
induced stresses by meters installed
in boreholes. The models provided a
good qualitative impression of the
behaviour of the roadways, with
"steel" arch deformation and
prominent floor lift being evident at
the higher loads, but not all the
aspects of behaviour were modelled
in a quantitative way, due mainly to
the problems of imitating rock
structure accurately in small models
of scale 1 to 150. The vertical
closure measurements at "site 3", and
the test rig used, are shown in
figures 3 and 4. The same rig was
used in later tests with a
geometrical scale of 1 to 100, in
which rock bolt modelling was used,
as reported by Mullins (1984).

4. Conclusions.

The obvious visual correspondence
between the models and field sites is
perhaps the most positive result, but
on a more quantitative basis the fact
that the modelling system could be
applied to scales of 1 in 150,
1 in 100 and 1 in 75 with comparable
results was most important. Difficulty
has been experienced in modelling rock
bolts, and particularly in ensuring that
they are properly mounted in the model
rock, but the future envisaged use of
scales of 1 in 50, and 1 in 25 for

detailed work, should enable this problem to be overcome. Methods of measuring the strength of rock and model materials have been developed so that reliable strength values can be obtained even from fragmented samples. Suitable strength scale factors can be chosen, which are not wholly dependent on the geometrical scale, enabling the same situation to be modelled in the same test rig with different strength scales, if this is desired. The results of early tests emphasised the importance of modelling rock structure, in addition

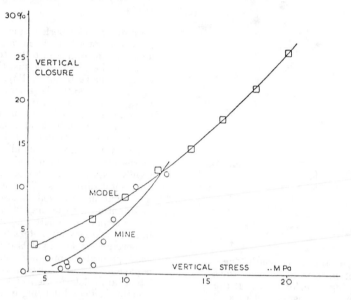

FIGURE 3. VERTICAL CLOSURE AT SITE 3

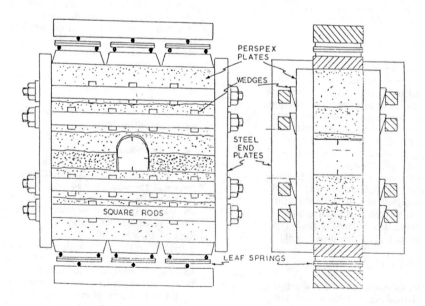

FIGURE 4. DIAGRAM OF 1 TO 150 TEST RIG

to strength and stress, this being particularly true of bedding planes. The modelling system devised should enable site specific models to be constructed and analysed quantitatively, and so reduce the extent of field trials necessary.

Acknowledgements.

The author would like to gratefully acknowledge the help of British Coal, (North Yorkshire), and MRDE, a number of research and project students, especially M. Hanif, A.A. Rayan, R. Krishna and D.R. Mullins, and colleagues of the Department of Mining and Mineral Engineering, University of Leeds.

References.

Brook, N. 1982. Small scale brittle model studies of mine roadway deformation. Proc. symposium on strata mechanics, 184-189. Newcastle-Upon-Tyne.

Brook, N. 1985. The equivalent core diameter method of size and shape correction in point load testing. Int. J. Rock Mech. Min.Sci. Geomech. Abstr. 22. 61-70.

Brook, N. 1987. Fluid transport of coarse solids. Mining Science and Technology. 5. 197-217.

Brook, N. 1990. The disc test. LUMA (Leeds University Mining Association). Ed. P.A. Dowd. 49-55.

Hanif, M. 1974. Support and deformation of underground openings. Ph.D. Thesis. University of Leeds.

Hobbs, D.W. 1966. Scale model studies of strata movement around mine roadways. Int. J. Rock. Mech. Min. Sci. 3, 101-127.

Mullins, D.R. 1984. Physical modelling of underground support systems. LUMA (Leeds University Mining Association). Ed. P.A. Dowd. 36-41.

Rayan, A.A. 1980. Application of rock strength tests to physical models. Ph.D. Thesis University of Leeds.

Effects of Geomechanics on Mine Design, Kidybiński & Dubiński (eds) © 1992 Balkema, Rotterdam. ISBN 90 5410 040 0

Energy analysis during compression tests of rock samples

D. Krzysztoń
Academy of Mining and Metallurgy, Cracow, Poland

ABSTRACT: Uniaxial compression tests on coarse-grained sandstone samples carried out in a stiff testing machine have shown that the stress-strain curves are not smooth on account of the local maxima and drops of stresses. According to the investigations performed by Minh (1989) the stress drops are caused by new cracks appearing in the rock sample. In this work Minh's hypothesis for the determination of the fracture surface energy consumed for the creation of new cracks in the pre-failure part of stress-strain relation was applied. An introduction of the new expression in the energy balance does not globally change the values of energy indexes characterizing the rock burst susceptibility and which, contrary to expectation, remain greater for water saturated than for dry samples.

1 INTRODUCTION

Experimental investigations carried out in a stiff testing machine on the properties of dry and water saturated samples of fine- and coarse-grained sandstones have shown considerable differences in the form of stress-strain relations (Krzysztoń 1989a). The rock samples were compressed at three different strain rates: $\dot{\varepsilon} = 2.10^{-6}$, 2.10^{-5}, 2.10^{-4} s^{-1}. At each of the strain rates, 4 - 6 dry and water saturated samples were tested. In Figure 1 two dependences between stress and longitudinal strain for both dry and water saturated samples of fine-grained sandstone tested at different sample strain rates are shown. These dependences have the form approximating a right-angled triangle with the hypotenuse overlaping the strain axis; for dry samples the pre-failure as well as post-failure parts of the stress-strain relations are straight lines, however for water saturated samples a small consolidation of material appears in the pre-failure part of the stress-strain relation, which manifests itself by the large increase in strain at small increase in stress. The pre-failure parts of water saturated samples have smaller angles of inclination than of dry samples, on the other hand the post-failure parts of dry and water saturated samples are parallel. The compressive strengths of dry samples are about twice as large as those of water saturated samples, but a difference in the strain values corresponding to ultimate stress for dry and water saturated samples does not appear, opposite to the Rebinder effect (Gustkiewicz 1987).

The form of dependences between stress and longitudinal strain for coarse-grained sandstone samples corresponds to the form observed in the investigation of different rocks (Wawersik and Fairhurst 1979).

The chosen stress-strain relations of dry and water saturated samples, obtained at the applied sample strain rates, are shown in Figure 2. Stress-strain curves are at first turned by convexity towards the strain axis, then there is a linear dependence between stress and strain, and before reaching the ultimate strength curves are turned by their convexity to the stress axis. For some samples the local stress maxima followed by stress drops were observed. It is distinctive that pre-failure parts of stress-strain relations for the dry and water saturated samples often overlap and a difference appears only in the values of the critical stress. The post-failure parts of the dry and water saturated samples are almost parallel and in the final stage the unstable failure of the sample manifesting by violent drop of stress to zero, at almost constant strain, takes place.

The relations between stress and

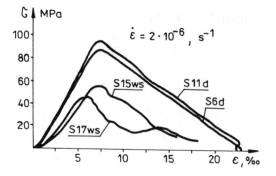

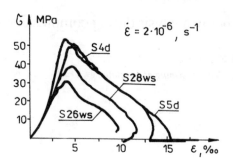

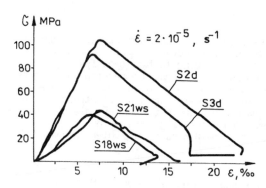

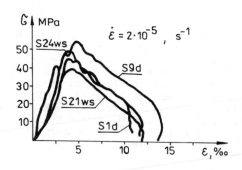

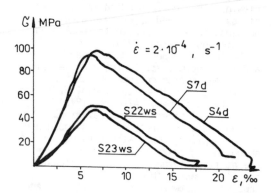

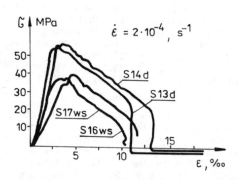

Fig.1 Stress as a function of longitudinal strain for dry and water saturated samples of fine-grained sandstone investigated at different sample strain rates

Fig.2 Stress as a function of longitudinal strain for dry and water saturated samples of coarse-grained sandstone investigated at different sample strain rates

longitudinal strain for fine-grained sandstone samples have the form corresponding to the case of stable destruction of a rock sample. The area contained between the stress-strain curve and the strain axis (Fig.3) determines the specific energy of longitudinal strain of a sample. This area was divided into two areas: under the increasing and decreasing stress-strain relation. Next the area under the

increasing part was divided into subareas corresponding to the elastic and irreversible strains, in relation to the ultimate strength of the sample (Gustkiewicz 1987). The areas of distinguished surfaces are the measure of the specific energy within particular ranges of sample strain. The introduced division of the area contained between the stress-strain curve and the strain axis corresponds to the division of the total

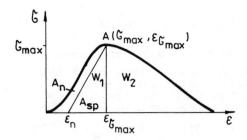

Fig.3 Division of the specific energy of longitudinal strain to component energies

energy W to the energy W_1 needed for reaching the strength limit and to the energy W_2 of stable destruction of the sample, where $W = W_1 + W_2$. The energy at strength limit W_1 is divided into the energies of elastic A_{el} and irreversible A_n strains. For determination of particular energies, the corresponding areas were planimetered and the values obtained were multiplied by the unit specific energy resulting from the assumed scales on the stress and strain axes. The introduced division of the longitudinal strain energy in the process of sample failure was applied to the calculation of the energy indexes: of rock burst $WET = A_{el}/A_n$, of hazard burst $WZT = A_{el}/W_1$ and of burst softening $WOT = A_{el}/W_2$. In the case of fine-grained sandstone, the water saturated samples had smaller values of specific energies within particular ranges of strain and smaller values of energy indexes than the dry ones. For coarse-grained sandstone the smaller values of particular energies were also obtained for water saturated samples than for dry ones but small differences in relative values caused that the values of energy indexes were greater for water saturated samples than for dry ones which was contrary to expectation. Therefore, the results obtained on the samples of coarse-grained sandstone had to be reconsidered.

2 ENERGY BALANCE DURING THE LOCAL DROP OF STRESS

Uniaxial compression tests carried out in a stiff testing machine proved that the stress-strain curves gained during tests

at constant strain rate were not wholly smooth because of local maxima of stress ("peaks") which appeared on the pre-failure part. At each local maximum the stress decreased abruptly and the sample lost to a certain degree its load-bearing capacity when the axial strain remained unchanged. The stress drops for the tested samples (sandstone, marl, limestone, granite, basalt) took place mostly at 80 - 90% of maximum stress (Minh 1989). A detail observation revealed that some of the stress drops took place simultaneously with the appearance of a new crack that could be seen by eye, on the sample side surface or could be registered by the increase of accoustic emission activity. In a way the amplitude of stress drops corresponded with crack size, the longer the new crack the greater the stress drop marked on the stress-strain curve. In the post-failure part the stress drops as well as plastic strains took place (Fig.4).

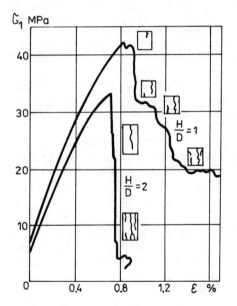

Fig.4 Progressive propagation of macro-cracks during compression of marl (after Minh 1989)

The level of the elastic strain energy stored in the compressed rock volume is related to the value of stress, hence the drop of stress indicates the drop of the stored energy level and its dissipation. The irreversible strains at compression are the results of cracking, slidings of crack interfaces and plastic flows. Two kinds of dissipated energy are connected with these phenomena. One of them is the fracture surface energy that is needed to create a new crack surface, the other one is the plastic energy consumed on generation of relative slidings and crystal translations.

The energy balance equation for the local stress drop has the following form (Minh 1989):

$$\frac{\partial W}{\partial t} = \frac{\partial U}{\partial t} + \frac{\partial S}{\partial t} + \frac{\partial P}{\partial t} + \frac{\partial C}{\partial t} \quad (1)$$

where:

- W - externally supplied energy
- U - elastic strain stored energy
- S - fracture surface energy
- P - plastic energy
- C - energy dissipated by acoustic emission,etc.

The stress drops recorded during compression tests took place without supplying external energy and at the axial strain almost unchanged. These mean the zero increments of external and plastic energies. Then the equation (1) reduces to the equation given in the increment form:

$$-\Delta U = \Delta S + \Delta C \quad (2)$$

Introducing the specific energy (per unit rock volume) and denoting the elastic strain energy, fracture surface energy and energy of acoustic emission by E_e, E_s, E_a, respectively, the energy transfer during each drop of stress can be expressed in the form:

$$-\Delta E_e = \Delta E_s + \Delta E_a \quad (3)$$

Assuming that the energy dissipated on the acoustic emission is relatively small (c. 2% of total value) then:

$$-\Delta E_e = \Delta E_s \quad (4)$$

Hence, it results that during the stress drop almost the whole released energy is transformed into the fracture surface energy, in other words, is consumed on the creation of a new surface.

3 DETERMINATION OF THE FRACTURE SURFACE ENERGY

The fracture surface energy during each stress drop can be calculated from the axial stress as follows:

$$\Delta E_e = (\sigma_1^2 - \sigma_2^2)/2E \quad (5)$$

where σ_1 and σ_2 denote the axial stress before and after stress drop, E is the longitudinal modulus of elasticity.

Hence the fracture surface energy is determined by the formula:

$$\Delta E_s = -\Delta E_e = \Delta\sigma \cdot \sigma_m/E \quad (6)$$

where $\Delta\sigma$ is the amplitude of stress drop, σ_m is the average axial stress.

Having determined from the stress-strain curve the stress which causes the fracture, the amplitude of stress drop and the modulus of longitudinal elasticity, it is possible to calculate the fracture surface energy.

The release of the strain energy in the pre-failure part can be determined analytically or graphically (Fig.5). The stress-strain curve behaves linearly up to certain stress σ_e. Above this value the stress drops as a result of progressive cracking. The rock volume undergoes a failure at maximum value of stress. In scheme (Fig.5) the deformation modulus E_o and the elastic modulus E_e obtained at sample unloading are marked. When stress drops are not observed then the maximum value of stress is in the point σ_c. In the case of stress drop occurrence the apparent peak stress $\sigma_c{'}$ is introduced. This stress is determined by adding the sum of stress drops to the real peak stress value (Fig.5):

$$\sigma_c{'} = \sigma_c + \sum_{i=1}^{n} \Delta\sigma_i \quad (7)$$

The total energy released in the pre-failure part is the sum of energies released during each stress drop (formula 6) and is equal to:

$$E_{sb} = \frac{1}{E} \sum_{i=1}^{n} \Delta\sigma_i \cdot \sigma_{im} \quad (8)$$

The average stress at which stress drops appear can be approximately expressed by formula:

$$\sigma_m = (\sigma_e + \sigma_c)/2 \cong \sigma_c \quad (9)$$

and then the pre-failure released energy used for the creation of a new surface is determined by the formula:

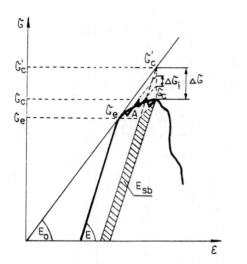

Fig.5 The pre-failure strain energy release (after Minh 1989)

$$E_{sb} = \Delta G \cdot G_c / E \qquad (10)$$

According to formula (10) the elastic energy released in the pre-failure part on the creation of a new surface for the coarse-grained sandstone samples was calculated. This energy is presented in approximation by the shaded area in Fig.5. In comparison with the division of specific energy of longitudinal strains to component energies (Fig.3) it can be assumed that the elastic energy accumulated in the sample E_{eb} is equal to the elastic energy, previously denoted by A_{el}; however the energy of irreversible strains in the pre-failure part E_{pb} is equal to the subtraction of the previously denoted energy A_n and the value of the elastic energy used for the creation of a new surface E_{sb}, i.e. $E_{pb} = A_n - E_{sb}$. The energy released in the post-failure part of stress-strain curve E_{pp} is equal to the energy of stable destruction of the sample W_2.

The determination of the fracture surface energy for dry and water saturated samples is given in Tables 1 and 2 where the energies determined by the planimeter method, i.e. the energy of elastic strain stored in the sample E_{eb} as well as the energies of irreversible (plastic) strain in the pre-failure E_{pb} and in the post-failure E_{pp} parts of the investigated sample stress-strain relations are given.

4 DETERMINATION OF ENERGY INDEXES

For determination of the rock burst susceptibility some indexes: bump energy index WET and brittleness factor K_e named also bump hazard index WZT were introduced. These indexes were determined on the basis of the pre-failure stress-strain curve. Considering that the burst rock susceptibility depends on the energy released in the post-failure part of stress-strain relation, the index $\lambda = M/E$ was used. This index depends on the slopes of inclination of the tangents to the post-failure (M) and pre-failure (E) parts of the stress-strain relation. The tangents are drawn at the values of stresses equal to 50% of the maximum stress. The index λ is applicable when the stress-strain curve can be approximated by straight lines. For most rocks the post-failure behaviour is not linear. Then the index called the energy ratio ER = E_{eb}/E_p is recommended. It exhibits the same tendency as WET and λ but is determined on the basis of the whole process of rock destruction.

The energy indexes calculated for the investigated samples of coarse-grained sandstone are given in Tables 1 and 2. The average values of these indexes calculated for the applied sample strain rates are presented in Figure 6. Here it is seen that the indexes determined on the basis of the pre-failure part (WET and WZT) of the stress-strain relation similarly characterize a bump hazard of the samples investigated. Also the indexes determined from the pre- and post-failure relations (λ and ER) heve the same changes tendency. It is important to notice that the water saturation does not decrease the rock burst susceptibility as it was observed on the sample of fine-grained sandstone. All energy indexes are greater for water saturated than for dry samples.

It is a well-known fact that the influence of water on rock properties depends on the structure of a given rock, i.e. on porosity, fissures, graining and the directional distributions of these defects (Alm 1982). Water saturation of fine- and coarse-grained sandstone samples caused the decrease in the compressive strength. However, the longitudinal strains of dry and water saturated samples have approximately the same values. This means that the Rebinder effect takes place only partially. This effect is the result of physical action of water on rock. The greater is the surface of micro-pores the greater is this action. In the coarse-grained sandstone the macro-pores with small developed surface

Table 1. Determination of the fracture surface energy and energy indexes in the failure process of dry samples

No.	ε̇ s⁻¹	No.of sample	G_c MPa	G_e MPa	%G_c	G_m	ΔG MPa	G_c'	E GPa	E_{sb}	E_{eb}	A_n kJ/m³	E_{pb}	E_{pp}	M GPa	WET	WZT	λ	ER
1.	10^{-6}	2	48	35	73	41	0.15	48	17.0	0.37	64	47	47	241	3.6	1.36	0.58	0.21	0.26
2.		3	60	38	64	49	–	–	22.7	–	80	23	23	328	3.6	3.50	0.78	0.16	0.24
3.		4	54	35	65	44	–	–	22.7	–	65	17	17	319	3.6	3.93	0.80	0.16	0.20
4.		5	51	38	74	45	0.25	52	20.5	0.54	70	38	37	250	3.5	1.86	0.65	0.17	0.28
5.		6	42	19	45	30	2.23	44	15.1	0.46	57	38	38	161	3.6	1.49	0.60	0.24	0.35
Average values			51	33	64	42	–	–	19.6	–	67	33	–	260	3.6	2.43	0.68	0.19	0.27
6.	10^{-5}	1	50	30	61	40	4.45	54	15.7	11.31	66	48	37	218	4.2	1.37	0.58	0.27	0.30
7.		7	43	32	74	38	1.93	45	17.7	4.13	85	29	25	165	3.9	2.93	0.75	0.22	0.51
8.		8	59	35	59	47	1.53	61	22.7	3.17	56	75	72	395	3.6	0.75	0.43	0.16	0.14
9.		9	55	37	67	46	1.59	56	20.5	3.54	71	56	52	316	3.5	1.28	0.56	0.17	0.23
10.		10	45	27	60	36	1.66	47	15.7	3.82	69	51	47	229	3.8	1.35	0.57	0.24	0.30
Average values			50	32	64	41	–	–	18.5	–	69	52	–	265	3.8	1.54	0.58	0.21	0.30
11.	10^{-4}	11	64	32	50	48	–	–	21.6	–	101	57	57	457	3.4	1.76	0.64	0.16	0.22
12.		12	51	27	52	39	0.32	52	19.9	0.63	67	58	57	289	3.6	1.16	0.54	0.18	0.23
13.		13	55	37	66	46	0.26	55	33.5	0.36	44	45	45	309	6.7	0.97	0.49	0.20	0.14
14.		14	57	41	71	49	–	–	29.9	–	59	34	34	366	3.6	1.75	0.64	0.12	0.16
15.		15	40	24	60	32	3.50	43	14.0	8.03	54	61	53	150	3.9	0.88	0.47	0.28	0.36
Average values			53	32	60	43	–	–	23.8	–	65	51	–	314	4.2	1.30	0.56	0.19	0.22

Table 2. Determination of the fracture surface energy and energy indexes in the failure process of water saturated samples

No.	ε̇ (s⁻¹)	No. of sample	G_c (MPa)	G_e (MPa)	%G_c	G_m	ΔG (MPa)	G_c'	E (GPa)	E_{sb} (GPa)	E_{eb} (kJ/m³)	A_n (kJ/m³)	E_{pb}	E_{pp}	M (GPa)	WET	WZT	λ	ER
1.	10^{-6}	26	32	22	69	27	1.11	33	11.7	2.60	41	15	12	106	3.4	2.74	0.73	0.29	0.39
2.		27	38	25	65	32	0.31	39	9.2	1.07	79	26	25	148	3.9	3.05	0.75	0.42	0.54
3.		28	39	25	66	32	–	–	12.7	–	57	24	24	163	3.6	2.35	0.70	0.28	0.35
4.		29	25	14	59	19	2.89	27	10.4	5.40	28	34	28	69	6.0	0.85	0.46	0.58	0.35
5.		31	30	14	47	22	3.18	33	8.5	8.35	54	26	18	91	5.0	2.02	0.67	0.59	0.59
Average values			33	20	61	26	–	–	10.5	–	51	25	21	115	4.4	2.20	0.66	0.43	0.44
6.	10^{-5}	21	40	29	72	34	–	–	17.0	–	47	25	25	193	3.6	1.90	0.66	0.21	0.24
7.		22	35	25	74	30	–	–	11.3	–	52	11	11	124	4.2	4.68	0.82	0.37	0.42
8.		23	31	15	47	23	0.64	32	8.9	1.65	53	22	20	110	4.6	2.43	0.71	0.52	0.48
9.		24	46	27	59	37	–	–	18.5	–	55	13	13	240	3.7	4.12	0.80	0.20	0.23
10.		25	51	31	62	41	–	–	17.3	–	76	50	50	285	3.4	1.53	0.60	0.20	0.27
Average values			41	25	63	33	–	–	14.6	–	56	24	24	190	3.9	2.93	0.72	0.30	0.33
11.	10^{-4}	16	37	24	64	30	–	–	16.3	–	42	14	14	171	3.4	3.01	0.75	0.21	0.29
12.		17	40	27	67	33	1.58	41	13.5	3.89	59	36	32	171	4.1	1.62	0.62	0.30	0.34
13.		18	32	22	70	27	–	–	13.5	–	36	30	30	99	2.3	1.20	0.54	0.17	0.36
14.		19	45	29	64	37	0.96	46	15.1	2.35	71	35	33	262	3.5	2.00	0.67	0.23	0.27
15.		20	55	29	52	42	0.64	56	17.7	1.52	81	55	54	324	3.4	1.47	0.60	0.19	0.25
16.		30	23	17	75	20	–	–	16.7	–	25	7	7	98	5.7	3.68	0.79	0.34	0.26
Average values			39	25	65	31	–	–	15.5	–	52	30	28	187	3.7	2.16	0.66	0.24	0.29

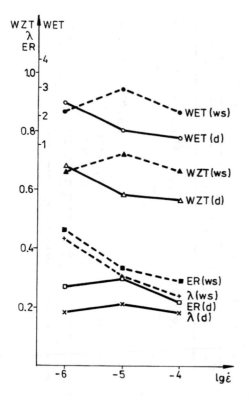

Fig.6 Average values of energy indexes at the sample strain rates

prevail, hence the physical action of water on rock properties can be smaller than in the case of a fine-grained sandstone.

The determination of energy indexes (excluding λ) was carried out on the basis of the percentage fraction of the elastic energy in the energies within particular ranges of the longitudinal strain of a sample. The transverse strain in uniaxial compression does not take part in energy analysis. However, from the graphs of the relations between longitudinal stress and transverse strain it results that the samples of coarse-grained sandstone undergo greater deformations than the samples of fine-grained sandstone (Krzysztoń 1989a). This must have an influence on the results obtained which has been proved in the classification of dry and water saturated samples of fine- and coarse-grained sandstones (Dzierwa 1990). An application of the new rock classification method (Turk and Dearman 1985) based, among other things, on Poisson´s ratio has shown that water saturated samples of fine- as well as coarse-grained sandstones are qualified to the classes of smaller strength and greater

deformability as comparised with their dry counterparts.

5 CONCLUSION

In the energy analysis of the progressive failure of rock samples the fracture surface energy used for the creation of new cracks in the pre-failure part of the stress-strain curve should be taken into consideration. This energy can be determined analytically or graphically. For the exact graphical determination of the fracture surface energy the compression tests should be carried out at loading and unloading of rock samples. In the case of the explosive failure of a rock sample the energy balance comprises also the kinetic energy, the values of which - determined directly from the energy equation - characterizes the intensity of a rock burst.

REFERENCES

Alm,O.1982. The effect of water on the mechanical properties of granitic rocks at high pressures and temperatures.Proceedings,Twenty-Third Symposium on Rock Mechanics.Berkeley,California,August 25-27,261-269.
Dzierwa,G.1990. Badanie mechanicznych własności skał nasyconych cieczą.Praca dyplomowa.Instytut Geomechaniki Górniczej, Akademia Górniczo-Hutnicza,Kraków.
Fras,S.1987. Określenie charakterystycznych parametrów skał dla różnych prędkości odkształceń.Praca dyplomowa.Instytut Geomechaniki Górniczej,Akademia Górniczo-Hutnicza,Kraków.
Gustkiewicz,J.et al.1987. Wpływ wody na mechaniczne własności skał tąpiących. Sprawozdanie etapowe. Instytut Mechaniki Górotworu PAN,Kraków.
Kłeczek,Z.1985. Mechanika Górnicza.Skrypty Uczelniane,Nr 1000,Kraków.
Krzysztoń,D.1989a. Pre- and post-failure stress-strain characteristics of dry and wet sandstone samples. The Second International Conference on Mining and Metallurgical Engineering,Suez Canal University,Proceedings,March 20-22,119-149.
Krzysztoń,D.1989b. Badanie energii odkształcenia podłużnego suchych i mokrych próbek piaskowca. Zeszyty Naukowe AGH, Górnictwo,z.145,Kraków,215-230.
Minh,V.C.1989. Energy analysis of deformation and failure of rocks. Rozprawa habilitacyjna.Wydział Geologii Uniwersytetu Warszawskiego,Warszawa.
Postół,W.1990. Wyznaczenie energii właściwej odkształcenia podłużnego w procesie

niszczenia próbki skalnej. Praca dyplo-
mowa. Instytut Geomechaniki Górniczej,
Akademia Górniczo-Hutnicza,Kraków.

Stachura,M.1991. Bilans energetyczny przy
odkształceniu i niszczeniu próbki skalnej. Praca dyplomowa.Instytut Geomechaniki Górniczej,Akademia Górniczo-Hutnicza,Kraków.

Turk,N.and Dearman,W.R.1985. A new rock
classification method for design purposes. 26th Symposium on Rock Mechanics.
Rapid City,South Dakota,June 26-28,
81-88.

Wawersik,N.R.and Fairhurst,C.1979. A study
of brittle rock failure in laboratory
compression experiments. Int.J.Rock Mech.
Min.Sci.,Vol.7,No 6,613-631.

Effects of Geomechanics on Mine Design, Kidybiński & Dubiński (eds) © 1992 Balkema, Rotterdam. ISBN 90 5410 040 0

Inversion analysis of original stress condition of rock strata based on extensometric measurement results

J. Aldorf
Mining University Ostrava, Czechoslovakia

ABSTRACT: The paper discusses the algorithm of the sequence and method of inversion analysis of the primary state of stress in rock mass using data obtained by measurements with the help of multilevel extensometers around the underground excavations.

Observation of underground working by monitoring makes available a great number of data and informations on the behaviour of rock strata and roadway support (convergence development, disturbance zones, loading, stress and deformation of support etc.). These data can be employed not only to cover projection needs, but they should be applied especially for creation of a geomechanic model by which development of the whole system behaviour, critical condition achievement, support reinforcement need etc could be predicted. Thus conceived geomechanic model, however, requires a continuous precisioning updating of input parameters which could be achieved by inversion analysis methods (IA) or back analysis, usually by the same computing model. Data required for such approach can be obtained as follows:
  - by convergence measurement (on excavation section, or support)
  - by deformation measurement (on support)
  - by dynamometric measurement (on support)
  - by dynamometric measurements of support loading or stress within support
  - by extensometric measurement (dislocations of points within rock strata.
  By inversion calculation then shape and size of support loading pattern, size of original stress tensor components, parameters ofrock strata deformational features etc can be determined. In lit. ref. (1) programming systems ofinversion analysis for convergence, deformation and dynamometric measurements are described. In the paper presented here its author´s intention is to inform about a single model IA based on extensometric measurements (see Fig.1) which when compared with the other above-mentioned systems are advantageous as they make available information on behaviour of broader roadway surroundings. The application of multi-level extensometers additionally gives more precise solution results as it allows the use of very accurate differential dislocation values between pairs of extensometer heads installed in various levels.

  The elaborated calculation system named INVERZE 4 is based on the description of deformation condition in the surroundings of underground roadway of arbitrary shape (representable by means of conform representation) and enables a determination of size of original stress tensor components in rock strata.

$$S_1 = \gamma \cdot h$$

$$S_2 = K_b \cdot \gamma \cdot h \qquad (1)$$

$$S_2/S_1 = K_b$$

The solution of this task is based on application of Kolosoff-Muschelishvili method with utilization of Melenteyeff method for determination of coefficients of conform representation. Stresses which have been caused by roadway drivage and which participate in dislocation genesis can be described by an abbreviated formula as follows:

$$\sigma_\rho = S_1 k_1' (\rho, \vartheta) + S_2 k_2' (\rho, \vartheta)$$
$$\sigma_\vartheta = S_1 k_3' (\rho, \vartheta) + S_2 k_4' (\rho, \vartheta) \quad (2)$$

The magnitudes of functions k_1' up to k_4' can be found out in (2). The magnitude of radial dislocation of M-point (see Fig.1) can be determined by correlation (see below) and numeric integration. For representation of abscissa $u_{(M)}$ (i.e. $u_{(M)} = W(\varphi)$) in the plane S_1 (conform representation) an iterative procedure by means of Newton method is applied

$$u_{(M)} = \sum^{2\nu} \int_{\rho_i}^{\rho_{i+1}} \frac{1}{E} \left[\sigma_\rho(\rho, \bar{\vartheta}_1)(1 - \mu^2) - \sigma_\vartheta(\rho, \bar{\vartheta}_1)(\mu + \mu^2) \right] \cdot \frac{\partial /w(\rho, \bar{\vartheta})}{\partial \rho} \cdot d\rho \quad (3)$$

The upper limit of integral (3) is a certain point $M(R_1, 0)$ in which the difference between additional and original stress components does not exceed 1% of original stress. In this point a virtually non-existent drivage caused dislocation can be experted.

If we write out the equation (3) for adjacent points M_1, M_2.

Then after an adjustment the following can be obtained:

$$\Delta u = u_{(M_1)} - u_{(M_2)} = \frac{1}{E} \left[S_1 \{ (1 - \mu^2) A - (\mu + \mu^2) C \} + S_2 \{ (1 - \mu^2) B - (\mu + \mu^2) D \} \right] \quad (4)$$

where the functions A, B, C, D represent integrals from the equation (3). If we designate

$\Delta \vec{u}^* = (\Delta u_1^*, \dots \Delta u_k^*)$ - vector of measured values

$\Delta u = (\Delta u_1, \dots \Delta u_k)$ - vector of calculated values

of pairs of measuring points (Fig.1) the unknown values of S_1, S_2 can be determined by a least squre method, i.e. by seeking a function minimum

$$f = \sum (\Delta \vec{u}_i - \Delta \vec{u}_i^*)^2$$

Out of extreme conditions the following results after adjustment

from the equations (4)

$$S_2 = K_b \gamma h = \frac{(\sum_1^k \Delta u_j^* L_2^j)(\sum_1^k L_1^{j2}) - (\sum_1^k \Delta u_j^* L_1^j)(\sum_1^k L_2^j \cdot L_1^j)}{\sum_1^k L_2^{j2} \cdot \sum_1^k L_1^{j2} - \sum_1^k (L_2^j \cdot L_1^j)^2}$$

$$S_1 = \gamma h = \frac{\sum_1^k \Delta u_j^* \cdot L_1^j - S_2 (\sum_1^k L_2^j \cdot L_1^j)}{\sum_1^k L_1^{j2}}$$

$$L_1^j = \frac{1}{E} \left[(1 - \mu^2) A_j - (\mu + \mu^2) C_j \right]$$

$$L_2^j = \frac{1}{E} \left[(1 - \mu^2) B_j - (\mu + \mu^2) D_j \right] \quad (5)$$

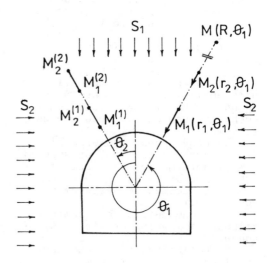

Fig. 1

In this way a computing program INVERZE 4 has been compiled which brings a solution of the above-mentioned problem and determines the values of original stress tensor components of rock strata S_1, S_2 and of lateral stress coefficients $K_b = S_2/S_1$.

REFERENCES

Aldorf, J., Hrubešová, E., Kořínek, R.: Features and methods of monitoring observation and inversion analysis applied on stability of shafts of Frenštát Colliery. TEZ VOKD 1/1981

Final report of VÚ SPZV II-6-1/04. 05 entitled "Stability and support of long roadways in a ggravated underground natural conditions. Mining University VŠB Ostrava, 1990

Effects of Geomechanics on Mine Design, Kidybiński & Dubiński (eds) © 1992 Balkema, Rotterdam. ISBN 90 5410 040 0

The model of structural mechanics and its theoretical basis for designing strata control technique in longwall panels

Song Zhenqi, Liu Yixue & Jiang Jinquan
The Research Centre for Strata Mechanics, Shandong Institute of Mining and Technology, People's Republic of China

ABSTRACT, The correct establishment of a model of structural mechanics is a prerequisite for designing strata control technique in the face area and the entries around the face. In the last ten years, the researches on theory and method for establishing the model taken aim at fixed quantity and directed against the local seam conditions have gained great successes. This paper introduces some achievements in the researches accomplished by the Research Center for Strata Mechanics, Shandong Institute of Mining and Technology, which include the theory for establishing the model, the structural parameters, and the method for determining the stress distribution etc. The achievements in the researches strive to establish the model developed from "qualitative stage" to "quantitative stage" directed against local seam conditions.

1. THE THEORETICAL BASIS FOR ESTABLISHING "MODEL"

Through the researches in the last many years, specially, the field investigations in the last ten years, authors put forward a new theory about the abutment pressure distribution, that is, "the theory of two stress fields", which provides a basis for establishing the structural model of strata control and determining some structural parameters. Summary of the theory is as follows.

1.1 In overburden strata, the strata influenced, obviously, the ground pressure distribution and the ground pressure manifestation by their movements are limited. Generally, this portion of strata consists of "the immediate roof" (A in Fig.1) and "the main roof" (B in Fig.1), in which the immediate roof strata have caved in the gob area and have not kept throughout the connexion transferred lateral forces to the solid coal in front of face line and the fragments in gob area. The main roof consists of the transferred rock beams (rock plates) moved and influenced obviously the ground pressure manifestation [1]. The influence of strata above the main roof on the ground pressure manifestation can only be embodied by the movements of main roof.

1.2 Under the condition of mining depth enough, the abutment pressures transferred to solid coal and strata around a coal face during the main roof breaking will obviously divide into two portions (Fig.1). The zone designated by S1 is called "internal stress field". The abutment pressure distributed in this zone arises from the active force of moving main roof, and its features of distribution and change are determined by the weight of main roof beams and its moving development. The zone designated by S2 in Fig.1 is called "external stress field". The abutment pressure distributed in this zone arises

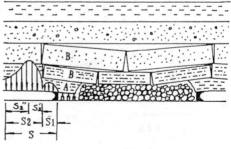

Fig.1

from the totality of overburden strata. The magnitude and distribution of abutment pressure depend on the active force arised from the mining depth or the total thickness of overburden strata, the strength of supporting solid coal (rock) and the boundary restriction conditions, etc. The relation between the extent of two stress fields and the overhanging roof area as well as the mining height can be approximated by linear equation for the same coal seam and constant mining height [2] [3].

1.3 The extent of stress distribution, the features of stress distribution and the magnitude of stress in each point within internal and external stress fields which develop with face advance and main roof movements are in changing throughout, uninterruptedly and regularly. After the main roof breaking and weighting, the main features of stress changing in two stress fields are that the stress in internal stress field will concentrate and shrink towards the face line with the face advance and main roof rotation. The stress distribution and peak stress position in external stress field will develop towards the front of face line. The changing will stop until the finishing of main roof movements, that is, it will enter a relatively stable state. During this period, the changing law in the direction of face advance and on two sides are shown in a and b in Fig. 2 respectively [2] [3].

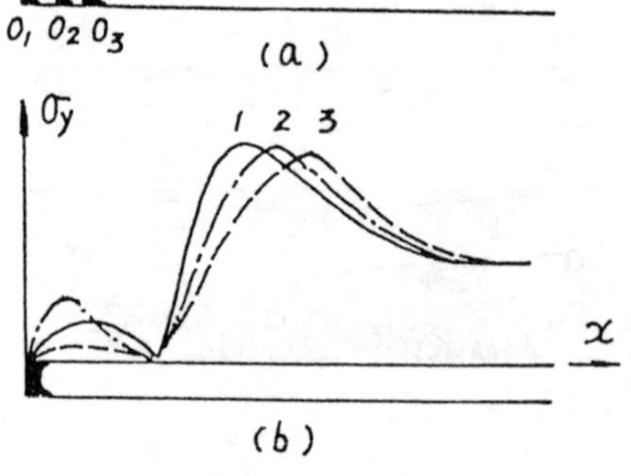

1.4 The stress changing in internal and external stress fields shows obvious periodicity. The changing law of periodicity in the direction of face advance is identical with periodic developing law of breaking and moving of rock beams (rock plates) formed the main roof, which is shown in Fig. 3. The number of changing period of side abutment pressure distribution and the changing developing process of each period is corresponding to the rock beam number formed the main roof, and the breaking and sagging of relevant rock beams (rock plates) respectively.

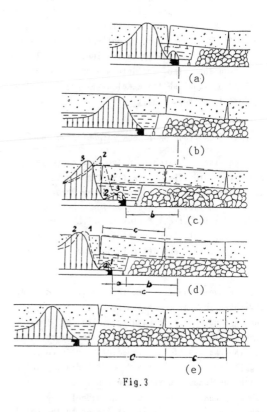

Fig. 3

1.5 The supporting resistance in a coal face (or supporting strength) has a great influence upon the magnitude and distribution of abutment pressure in internal stress field. This influence is realized by controlling the position and time of beginning of rotation (the distance from face line) after the main roof beam breaking.

Fig. 2

Obviously, the development of coal (rock) failure in internal stress field can be controlled by regulating the support resistance in order to achieve changing the boundary restriction condition of external stress field and controlling the features of abutment pressure distribution in external stress field including the extent of stress distribution and the depth of peak pressure from face line.

2. THE MODEL OF STRUCTURAL MECHANICS IN COAL FACE AND ITS APPLICATION TO DESIGNING STRATA CONTROL TECHNIQUE

Influenced by the mining depth, the mining height, the strength of coal, the strength of roof and floor, and the overhanging roof area after mining, the model of structural mechanics of main roof breaking and weighting for coal face can be summed into two basic patterns.

2.1 The model without internal stress field. The abutment pressure distribution is shown in Fig.4a. Generally, this model appears in smaller mining depth (less than 250~300m), smaller mining height, hard coal, hard roof and floor, and smaller overhanging roof area after mining. That is, coal (or rock) can not be failed in compression by the maximum abutment pressure.

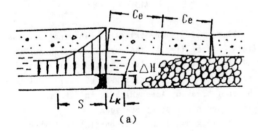

(a)

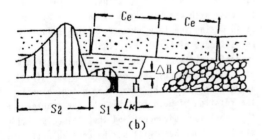

(b)

Fig.4

2.2 The model with internal stress field is shown in Fig.4b. Generally, the model appears in the depth of greater than 300~400m unless very hard coal seam (f<3~4). The model also appears in the depth of less than 150m for soft coal, roof and floor.

Under the condition of model without internal stress field, generally, the immediate roof breaking in advance is not considered during designing strata control. It is proven by researches and practices [4] that the required supporting strength P which keeps the broken main roof beam in equilibrium under a certain position state can be approximately expressed by "the characteristics equation", that is,

$$P = A + Ka \frac{\triangle Ha}{\triangle H}, \quad kpa \qquad (1)$$

In equation [1], $\triangle Ha$ is the roof convergence of coal face when the rock beam sags to the final position, i.e. it contacts with the rock fragments in gob area. In the condition of a certain mining height (h), a certain face width from face line to breaking line (Lk) and a certain bulking factor at the contact point between the rock beam and rock fragments (Ka=1.3~1.35), when the thickness of immediate roof (Mz) and the periodic roof weighting interval of main roof (Ce) are known, the magnitude of $\triangle Ha$ can approximately be solved by

$$\triangle Ha = \frac{[h - Mz(Ka - 1)]}{Ce} Lk \qquad (2)$$

In equation (1), Ka is a position state constant, (i.e. a force on the support exerted by main roof when the roof convergence reaches $\triangle Ha$). When the thickness of rock beam (Me), the volume weight (γe) and the periodic roof weighting interval (Ce) are known, the magnitude of Ka can be expressed by

$$Ka = \frac{Me \gamma eCe}{k \cdot Lk} \qquad (3)$$

where k is a factor reflecting the percentage of rock beam weight borne by the support. Generally, k=1.5~2.0, the higher the claimed position state to be controlled (the smaller the magnitude of $\triangle H$), the smaller the magnitude of k is.

In equation (1), A is the required supporting strength for controlling the immediate roof. When the thickness of immediate roof (Mz) and the volume weight (γz) are known, the magnitude of A can be determined by

$$A = Mz \gamma zFz \tag{4}$$

where Fz is the moment factor, according to the overhanging roof length (Ls) and the distance from active point of resultant on the support to the face line (Li), it can be determined by

$$Fz = \frac{Lk}{2Li}(1 + \frac{Ls}{Lk})^2 \tag{5}$$

In equation (1), $\triangle H$ is the claimed roof convergence to be controlled in coal face for safety in production. In the condition of this model, when the main roof is formed from a single rock beam. One obtains $\triangle H = \triangle Ha$ for the control scheme of "free deformation" (i.e. the support is in a free deformation state). In the condition of multiple rock beams, during the upper rock beam breaking, the lower rock beam contacts the upper rock beam as tight as possible in order to prevent the impact of upper rock beam. The magnitude of $\triangle H$ should approximate the minimum roof convergence ($\triangle Hmin$) measured in coal face.

Once the supporting strength P is determined, the supporting density (N) in coal face with hydraulic props can be calculated by

$$N = P/Rp \tag{6}$$

where Rp is effective bearing capacity of a hydraulic prop.

The yield resistance of a powered support (Rt) can be calculated by

$$Rt = SP, KN \tag{7}$$

where S is the roof area borne by a powered support (m^2).

Under the condition of structural mechanics model with internal stress field, the equation kept the structure in equilibrium can be simplified by

$$\int_0^{Sl} 6\,ydx + Rp = Qa + \frac{Qe}{2} \tag{8}$$

Where Qa and Qe are the active forces exerted by the immediate roof and the main roof respectively at the time of the formation of model, Rp is the supporting resistance (Rp = PLk).

The model as Fig.5 is used for determining the magnitude of abutment pressure ($\int_0^{Sl} 6\,ydx$) distributed in internal stress field (S1). After the stiffness (Go) and compressive value (Yo) of

solid coal are treated as linearity, the equation can be derived, that is

$$\int_0^{Sl} 6\,ydx = \frac{G0Y0S1}{6} \tag{9}$$

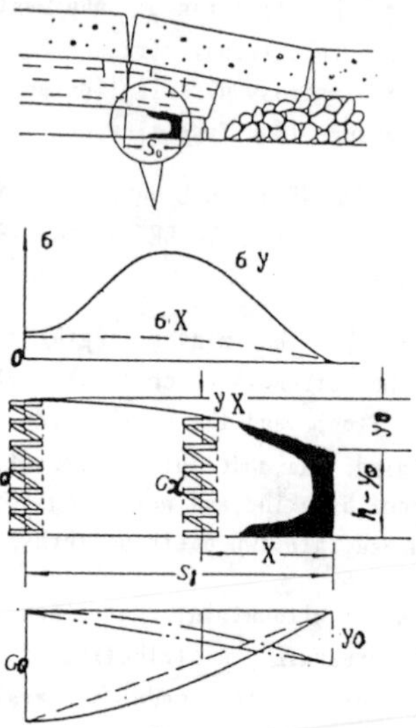

Fig.5

Thus, under the condition of this model, when the main roof is in corresponding position state $\triangle H$, the required supporting strength P is

$$P = (A + Ka\frac{\triangle Ha}{\triangle H}) - \frac{G0Y0S1}{6Lk} \tag{10}$$

Where S1 is the extent of internal stress field at the time of the rock beam breaking, Y0 is the compressive value of solid coal along the face line, G0 is the stiffness of solid coal near the roof breaking line which is in plastic strain state, and other signs are idem.

It is known from equation [10] that when the rock beam is located in a claimed position state $\triangle H$, the required supporting strength P will increase with the shrinkage of internal stress field during the face advance and the reduction in stiffness of solid coal. That is, if the support resistance is chosen from the magnitude of P calculated by the maximum stress field extent formed at the time of main roof breaking, the rotation of main roof will be inevitable once the coal face advances, and the claimed position state of rock beam will not be kept as well. The solid coal in internal stress field is failed and squeezed out, and the breakage of immediate roof is extended.

The stress located at distance x from the face line within internal stress field can be derived from the model as shown in Fig.5, that is

$$6_{yx} = \frac{G0Y0X(S1-x)}{S_1{}^2} \qquad (11)$$

The derivative of equation is $d6_{yx}/dx$, setting $d6_{yx}/dx=0$, the location of peak stress in internal stress field is $x=S1/2$. Substituting $x=S1/2$ into equation (11), the maximum value of stress in this location is

$$6_{max} = \frac{G0Y0}{4} \qquad (12)$$

Thus, the relation between the maximum stress in internal stress field and the support resistance established from the equilibrium condition as mentioned above can be expressed by

$$6_{max} = \frac{3Qe+6(Qa-Rp)}{4S1} \qquad (13)$$

It is known from equation (13) that the magnitude of support resistance performs an important function for controlling the ground pressure manifestation such as the concentrated stress in internal stress field, and solid coal breaking and squeezing out. This function will intensify with the face advance and the shrinkage of internal stress field (S1).

As mentioned above, under the condition of having internal stress field, the mechanics model for designing the roof control technique should be determined by the local roof conditions and the needs for controlling breakage and squeeze out of solid coal. When the immediate roof is very hard and the solid coal needs to be broken by ground pressure, the supporting resistance can be calculated by the model shown in Fig.6a. That is, the required supporting strength P can be calculated by substititing $\triangle H = \triangle Ha$ into equation /10/.

When the immediate roof is easy to be broken or the ground pressure is liable to impact, the supporting resistance should be determined by the model shown in Fig.6b. The required supporting strength P can be calculated by $S1=0$ (i.e. the main roof begins to rotate once the breaking line of main roof enters face area) and substituting

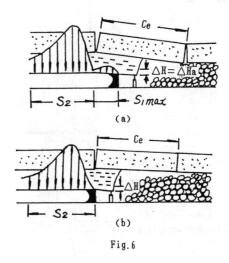

(a)

(b)

Fig.6

the position state to be controlled $\triangle H$ into equation (10).

3. THE MODEL OF STRUCTURAL MECHANICS OF GROUND PRESSURE IN THE ENTRIES AROUND COAL FACE AREA AND ITS APPLICATION TO DESIGNING STRATA CONTROL TECHNIQUE

It is proven by researches that the basic law of main roof breaking with face advance for the coal face unmined at its two sides in a flat coal seam is shown in Fig.7. The developing process of abutment pressures in the direction of face advance and on two sides is synchronous, and their distribution features and relevant parameters including the stress distribution in internal stress field and external stress field as well as the position of peak pressure, etc. are basically approximate. The developing law of abutment pressure is shown in Fig.8.

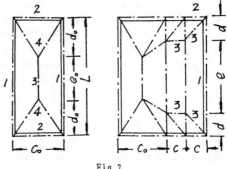

Fig.7

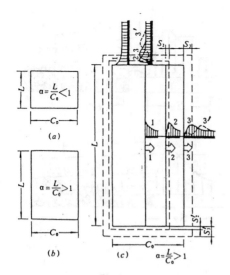

(a) $\alpha = \dfrac{L}{C_0} < 1$

(b) $\alpha = \dfrac{L}{C_0} > 1$

(c) $\alpha = \dfrac{L}{C_0} > 1$

Fig.8

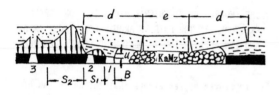

Fig.9

Thus, the model of structural mechanics of ground pressure in entries based on the finishing state of main roof weighting for the coal face with internal stress field is shown in Fig.9. Therefore, when the roof movement parameters of model are known, the reasonable location and time for driving the entries can be determined, and the deformation values of surrounding rock influenced by face extraction can be predicted. The entries to be maintained in gob area are located in the stress relief area over a long period of time after the roof movements have stabilized. The roof convergence u in entry caused by the main roof movements is the main portion of surrounding rock deformation, which is expressed by

$$u = \frac{S_1 + B}{d}[h - M_z(K_a - 1)] \qquad (14)$$

It has relation to the side span of main roof(d), the width of internal stress field(S_1), the width

of entry (B), the mining height (h), and the thickness of immediate roof and its bulking factor.

The reasonable location of entries is in internal stress field and is drived along the gob edge or along the gob edge left a small pillar. The reasonable time for driving the entries in internal stress field is at the time when the main roof movements have stabilized.

The supports in entries to be drived along the gob edge for controlling the strata movements should satisfy the following requests. Firstly, the value of support yield should correspond to the final stable state of main roof movements. Fig.9 shows that the main roof can not form the structure similar to an arch with three articulations in the direction of dip after it has moved remarkably, and the entries to be maintained in the gob will go through the whole process of overburden movements including the developing stage and the final stable stage. Therefore, the supports in the entries to be maintained in the gob, specially in the entries to be drived in internal stress field can not impose restrictions on the final position of main roof movements. Thus, the value of support yield in internal stress field should correspond to the final stable state of main roof movements. Secondly, the supporting resistance perpendicular to roof should maintain the stability of immediate roof. Considering the loss of bearing capacity of small pillar, the supporting resistance in the entries to be drived along the gob edge and along the gob edge left small pillar can be calculated by equation [4] as mentioned above.

In the condition of model without internal stress field, the solid coal is in elastic strain state, and $S_1 = 0$. Therefore, the entries to be maintained in the gob are easy to be maintained because the deformation in the entries caused by main roof movement is smaller. After stability of main roof movements, the entries to be drived along the gob edge are also reasonable.

4. DETERMINATION OF THE PRINCIPAL PARAMETERS IN A STRUCTURAL MODEL

In structural models of face and entries, the principal parameters are the thickness of immediate roof (M_z), the number of main roof beam and its thickness (M_e), the interval of main roof movements (Ce), the extents of internal and external stress fields (S1 and S2), and the side span of main roof (d).

4.1 The thickness of immediate roof (Mz)

The deductive method of caving condition, the caving condition of roof layer No. i is

$$Mi \leqslant h - \sum_{j=1}^{i-1} Mj(Ka-1) \qquad (15)$$

It is judged one by one from lower layer to upper layer if the roof has caved. The total thickness of caved strata is Mz.

The deductive method of position state, if the interval of measured periodic main roof weighting is Ce, and the roof convergence at the time of finishing of roof weighting is $\triangle$Ha, Mz can be determined by

$$Mz = \frac{h - Sa}{Ka - 1}, \quad Sa = \frac{c \cdot \triangle Ha}{Lk} \qquad (16)$$

The deductive method of **gravity** a supporting strength measured during the immediate roof moving alone is P', Mz can be determined by

$$Mz = \frac{p'}{\gamma \, zFz} \qquad (17)$$

4.2 Number, thickness and moving interval of main roof

The estimation of extent of main roof according to statistic experience, generally, the extent of roof strata required to be controlled is equal to$(6\sim8)$h, and minus the thickness of immediate roof is the extent of main roof.

The delimitation of composite rock beam according to simultaneous moving condition, when the upper rock layer j and the lower layer i composed a composite rock beam which can move simultaneously, the condition is

$$EiMi^2 > K^4 EjMj^2, \quad K = 1.15 \sim 1.25 \qquad (18)$$

thus, 2-3 lower composite rock beams can be delimited.

Determination of the number and the moving interval of main roof beams by actual measurement, the changing law of roof convergence and support resistance with face advance can be obtained by actual measurement. The unmber and the moving interval of main roof can be determined by the periodicity of changing law.

4.3 The extent of internal and external stress fields (S1 and S2)

According to the abutment pressure manifestation, and the relation between roof movements and abutment pressure distribution, the extent S1 and S2 can be determined by monitoring the roof-to-floor convergence in entries, and collecting shrinkage and expansion of internal and external stress fields. The monitoring can be achieved by using the advanced DKJ-1 coal mine ground pressure computer monitoring system in order to realize automatic monitoring and analys-ing by surface computers.

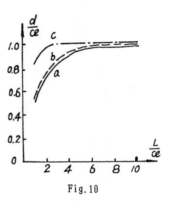

Fig. 10

4.4 The side span of main roof

It is proven by limiting analysis of main roof strata [4] that there is a certain relation between the side span (d) and the moving interval (Ce), which is shown in Fig. 10. When two sides are solid coal (curve a), one side is solid coal and another side is coal pillar (curve b), and L/Ce>5, one obtain d≈Ce. When one side is gob area (curve c) and L/Ce>2, one obtain d≈Ce. That is, if the face is long enough, the side span of main roof is equal to periodic moving interval.

REFERENCES

[1] Song Zhenqi and others. 1988. Applied Coal Mine Ground Control. The Publishing House of Chian University of Mining and Technology.

[2] Song Zhenqi and others. 1983. Auswirkung des tragenden Drucks Vor und nach dem Bruch der Gesteinsbalken -Anwenden bei der Abbauplan-

105

ung. 8th plenary Scientific Session
International Bureau of Strata
Mechanics／Essen／22-24.

[3] Son yang and Song Zhenqi. No.1, 1984.
     ....nal of Coal Mine. Abutment Pressure
     Manifestation in Coal faces and its Relation
     to Overburden Strata.

[4] Jiang Jinquan.No.1,1988.Journal of Shandong
     Institute of Mining and Technology. The
     Breaking Law of Plate Structure for Main
     Roof Strata in Coal Face.

Effects of Geomechanics on Mine Design, Kidybiński & Dubiński (eds) © 1992 Balkema, Rotterdam. ISBN 90 5410 040 0

Determination and prediction of geomechanical state of rock mass

I. M. Petukhov

Research Institute for Geomechanics and Mine Surveying (VNIMI), S. Petersburg, Russia

ABSTRACT: The paper emphasizes a need for continuous control of stress-strain state of rock mass and its gas-hydrodynamic condition that will enable to make a forecast of their changes at all stages of deposit mining. Methods and techniques for the safe mining are described.

INTRODUCTION

It should be noted from the outset that due to the development of rock mechanics as a science and its extended use in solving the problems concerning the safe, economical and cost-effective exploitation of mineral resources with due account of environmental control, work is currently progressing towards new methods and techniques for continuous control of stress-strain state and gas-dynamic condition of rock mass. As an example, consider the experience gained in prevention rockbursts and outbursts in coal deposit mining in the USSR /1/. It is of interest to note that for the last 15 years, as applied to all coal mines, a set of regional measures has been developed for control the coal seams prone to rock bursts and outbursts, including the advance mining of protective beds, degasification of gas-hazardous beds in the relief zones, regional wet treatment of coal strata within the whole level, pillarless mining of coal beds using a preliminary coal preparation. It follows that introduction into practice of this set of regional measures only at 40 coal mines contributed to avoid the occurrence of rockbursts and outbursts as well as to receive a pronounced economic return excluding the use of local preventive measures. It is significant that this experience obviously shows that gas-and rock pressure control

being used at all stages of deposit exploitation, makes it possible as if to reduce the actual mining depth twice or four times followed by favourable outcome /1/. There is no doubt that for proper gas-and rock pressure control one must timely determine the geomechanical parameters of rock mass and make a forecast of their changes at all stages of deposit mining.

DISCUSSION AND NEW OBSERVATIONS

Consider this matter taking into account the latest achievements in science on Earth including rock mechanics and geodynamics of rock mass which are progressing essentially in recent years in exploitation of deep-seated deposits.

It is known that the Earth's crust consists of two tens of plates continually interacting with each other. The continents and oceans "are floating" on these plates. As a result, force interaction of plates causes their breaking into megablocks, blocks of different dimensions and shape up to macrostructure of rocks detected in mining operations and in erection of underground structures.

As it is impossible to take into account the all variety of shapes and dimensions of rock mass components, their properties, interacting processes in solving specific problems in practical activities of a man, so the necessity arises in seeking a phenomenological approach based on the established regularities of stress distribution in the earth crust on the whole and in its individual sections with due account of averaged physico-mechanical and chemico-biological parameters of rock mass and its temperature.

It should be noted that a natural stress field in each section of the Earth represents a superposition of effects of planetary and tectonical fields of stress. Horizontal tectonic forces acting within the earth crust, apparently will be defined to a great extent by the processes which concern its increment in the regions of oceanic and terrestrial rift zones as well as by absorption within zones of subduction of oceanic plates under the continental plates /3/.

It follows from /4/, that when analyzing the stress state of the earth crust one can tentatively conclude that its individual sections (as to dimensions and depth) would be considered as having the following properties: a) elastic state of rock mass with absence of horizontal compression by active tectonic forces; b) elastic state of rock mass with presence of horizontal compression by active tectonic forces; c) ultimate strain state of rock mass at some depth from the earth surface with presence of horizontal compression by active tectonic forces; d) ultimate strain state of rock mass at overall depth of the earth crust with presence of horizontal compression by active tectonic forces. The mentioned work presents the methods for analysis of stress state of rock mass with reference to each individual section of the earth crust.

In general, to study the structure and stress state of rock mass within that or another deposit, as well as to find out to what section of the earth crust this deposit refers, one can use the method of geodynamic zoning of mineral resources /2/. With respect to this method, one carries out the following operations in a step-by-step manner

MAPPING OF BLOCK STRUCTURE OF ROCK MASS

Block structure of rock mass can be specified on the basis of the morphostructural analysis of crustal relief and, if possible, also the interjacent surfaces (floor of coal seam, deposit or another reference horizon). For the purpose of specifying the blocks of I-IV classes one can use, primarily, the maps on a 1 : 2000000 or 1 : 25000 scale. Blocks are specified by the principle from the general to the particular. Extension of the block network takes place gradually. The position of tectonic block boundaries is being revised in comparison with data obtained from satellite and aerial surveys, geological and geophysical prospecting as well as from meteorological surveying, etc. This operation is resulted in making-up a complete set of maps showing a block structure of a deposit.

STUDY OF ROCK PROPERTIES AND THE FACTORS RESPONSIBLE FOR ROCK FAILURE

It is known that this matter is of prime importance and is greatly essential to minerals mining. And there is no doubt now that an approach to solving the problem concerning the geomechanical state of rock mass at great depth, in particular, the exploitation of petroleum and gas deposits as well as the drilling of super-deep wells should be substantially revised. So, in this regard, in our country work has been currently under way on designing the high-pressure unit having the ability to simulate the field conditions as to stress and temperature up to the depth of 10 km – 15 km. This unit is designed for testing the physico-mechanical properties of rocks, for study the factors being the cause for their failure as well as for drilling equipment tests. The unit has resulted from a joint effort by two Institutes (the Perm Institute and VNIMI); now it is test-exploited at a special proving ground near the town of Perm. Testing specimens measure 260 mm by 1500 mm. The temperature can vary within 0 to 300°C The unit is 50 m in length; it is positioned in the mine shaft. Peak stress reaches the value of 500 MPa axially and 500 MPa - across the test specimen.

DETERMINATION OF TECTONICALLY STRAINED ZONES

Study the matters as to block interaction and various disturbances of rocks contributes to reveal the tectonically strained zones as well as the relieved ones. These zones first of all, are timed to those tectonically disturbed areas where fault has not yet revealed itself: on block structure maps such areas are denoted with dotted lines. Furthermore, tectonically strained zones and the relieved ones very often are caused by rock movements inside the rock mass along the available uneven weak planes (faults, tectonic disturbances), by existence of solid inclusions within the rock mass as well as by variations in thickness of individual layers, etc. To reveal the tectonically strained zones and the relieved ones, one can use data obtained from geomorphological, geological, seismological investigations, seismic prospecting as well as from study of phase-physical prpperties of coal seams, quality of coal, etc. Tectonically strained zones are plotted on maps of block structure and mine plans.

EVALUATION OF NATURAL STRESS STATE AND GAS-DYNAMIC CONDITION

On the results of study of geodynamic interaction of blocks one can ascertain the directed action and the calculated values of averaged principal stresses for the deposit (mine field) position. Using these values as boundary conditions one can calculate the principal stresses of rocks for a deposit on the whole and for each block individually in their native state (prior to initiation of mining operations).

Special form of boundary integral equations needed for a problem on system of interacting blocks with random conditions at contact surfaces, has been solved on a computer. The boundaries at which blocks interact, represent displacements, faults, ruptures, rift zones and intense jointing. They were formed principally, due to the vertical and horizontal mutual displacements of blocks or their combined movement in several directions. At the block contacts the stress and move

ment of rocks are intimately bound up with linear relationships. By using them, one can express the conditions of smooth contact, constant friction, complete cohesion of rocks, etc. This method enables to make stress calculations within block rock mass with random block interaction at the contact planes as well as to determine various elastic properties of the block themselves.

The method of boundary integral equations is used for analysis the stress state of block rock mass as a virgin one and also at any stage of deposit mining. The calculation results can be corrected, if necessary, in comparison with data obtained from full-scale measurements of stress at individual points of rock mass.

Based on the geodynamic zoning, it seems to be possible to carry out the flow investigations as well as to predict the properties of rock-collectors being under influence of non-uniform force fields. For this purpose we adopt a model of planar flow in any stratum with specified rate of flow, considering that a producing horizon is represented by quasiisotropic jointed-porous rocks with chaotic distribution of joint systems. In this case, ignoring the block structure of a deposit, the plan will be uniform in permeability, changes of which will show the manifestation of pure geodynamic factors. For flow prediction one can use the theoretical results combining the structural and flow characteristics of beds with stress state.

One draws a map of isolines of the permeability components in block structure. The values of permeability are given in its proportions conforming to the variant without reference to block structure. Due to impact of high tectonic stresses depending on the seepage properties of the medium its other characteristics will change. In particular, water and gas replacement takes place within the area of relieved stress. Maps of isolines of water-and gas saturation of a productive bed can be constructed.

EVALUATION OF STRESS STATE AND GAS-DYNAMIC CONDITION OF ROCK MASS AT DIFFERENT STAGES OF DEPOSIT EXPLOITATION

Having the structure and stress state of rock mass in a virgin condition as the reference point, one can make then the stress calculations with due account of mining operations based on the solution of spatial problem on stress distribution according to the programme PLANES /5/. In this regard one takes into consideration a complicated configuration of productive workings within the strata. For the purpose of minimizing an amount of calculations very often it is convenient to use the methods based on solving the plane problem: two programmes SHVARS and BLE have been developed /5/. There is a tendency to present the results of calculations in the form of "prognostic maps" for a deposit (mine field) on the whole or for its any sections (situations). Prognostic maps can be constructed for any period of exploitation of that or another object up to its complete extraction. The availability of prognostic maps offers an opportunity to analyze the geomechanical state of rock mass, its rockburst and outburst hazard within the exploited mine field or a group of fields, to choose the suitable methods for their mining as well as to verify the necessity for the use of protective measures in carrying out the safe, cost-effective and economical mining operations.

INSTRUMENTAL OBSERVATIONS FOR THE ASSESSMENT AND PREDICTION THE STATE OF ROCK MASS

Regional assessment is made on the basis of micro-seismological observations within a deposit or mine field by using the underground and surface pavillions. One determines the coordinates and energy of bumps in rock mass followed by construction of maps for regional prediction of rockburst hazard. Recently the successful attempts are made in forecasting heavy bumps (more than 10^6 J) at the site of their occurrence and with due account of time factor. Similar investigations

gain a prime significance in view of an idea concerning a theory to be ventured for bump-like deformation of rock mass as well as the methods and techniques for rock failure control being a further contributory factor to the safe and cost-effective mining the deposits prone to tectonic phenomena manifestation.

It should be noted that the stress state of rock mass is systematically analyzed followed by its prediction immediately at the site of mining operations by using a rapid analysis for forecasting rock bursts. This rapid analysis is based on records of electromagnetic and acoustic emission in rock mass. For this purpose a system VOLNA is used, now being in a serial production. It is expected that very soon in our country this method will find wide use at all coal,ore and non-metalliferous deposits prone to rock bursts.

CONCLUSIONS

It can be concluded from the above that for preliminary regional control of rockburst and outburst hazardous rock mass and, in general, for gas-and rock pressure control for the purpose of safe, cost-effective and economical exploitation of deposits, one should obtain timely a complete set of reliable data on structure, stress state and gas-dynamic condition of rock mass, physico-mechanical properties of rocks and desired changes of them. Such analysis of geomechanical state, as it was shown above, can be made by using the method of geodynamic zoning of mineral resources /2/, having found wide use in exploitation of hard and liquid deposits for the last ten years in our country.

REFERENCES

1. Petukhov,I.M. & Kuznetsov,V.P. 1991. Order of the change-over to regional control of coal seams prone to rockbursts and outbursts in designing and exploitation of deep mines. Leningrad:VNIMI.38 p.(in Rus.)
2. Petukhov,I.M.& Batugina,I.M.1990.

Geodynamic zoning of mineral
resources: Methodical instructions. Leningrad:VNIMI. 130 p.
(in Russian).
3. Petukhov, I.M. 1991. On nature
of horizontal forces in the
Earth's crust. Proc.,Lening-
rad:VNIMI.(In Russian).
4. Petukhov, I.M.& Sidorov, V.S.
1990. Estimation of principal
stress values within the compression zones of the Earth's
crust. Proc.,P.I. Leningrad:
VNIMI.p.144-150.
5. Methodical instructions for
stress calculations in the vicinity of productive workings.
1989. Leningrad:VNIMI. 39 p.
(in Russian).

Effects of Geomechanics on Mine Design, Kidybiński & Dubiński (eds) © 1992 Balkema, Rotterdam. ISBN 90 5410 040 0

Interpretation of the results of rock pressure upon tunnels lining at full-scale measurements at the lining stage construction

Nina N. Fotieva, A. S. Sammal & A. K. Petrenko
Tula Polytechnical Institute, Russia

ABSTRACT: The method of interpreting the data of full-scale measurements of pressure upon tunnel lining offering a possibility to evaluate the stress initial field characteristics in an intact massif is given in the paper. Examples of experimental-analytic design of lining constructed in stages are given.

The process of the lining being designed together with the rock massif as a single deformable system, the characteristics of the stress initial field in an intact massif: the main stress value N_1, the main initial stresses relation $\lambda = N_2/N_1$ and the initial stresses main axes inclination angle α to the vertical and the horizontal substantially influence the results being received, the results mentioned being applied as initial data. It is with those characteristics that the qualitative characteristics distributing normal and tangential contact stresses (loads) around lining cross-section perimeter and, therefore, stresses and forces in lining sections is to a large extent evaluated. Besides, the correcting multiplier α^* (Fotieva 1980) is introduced into the initial stresses values, which is called forth by the necessity of taking the lining construction lagging behind the face working and the non-linear rock deformation into account.

The circumstances indicated serve as the reason for attempts of creating tunnel lining design methods on the base of stresses measured in-situ which is particularly up-to-date for designing underground structures in tectonic active regions where initial stress field in the massif may substantially differ from the ones called forth by the weight of rocks lying above, both by the main stresses value and by its main axes direction.

Taking it into consideration the method of indirect evaluation of the λ, α massif characteristics necessary for designing and also a correcting multiplier α^* on the base of the results of full-scale measurements of normal pressure the mining working support and underground structures have been elaborated at Tula Polytechnical Institute (Fotieva & Bulychev 1980, 1981,1982). These characteristics evaluated on the basis of full-scale measurements don't depend upon the construction parameters and may be applied for the computing and designing another (different from the measurement object) lining in similar mining-geological conditions at a similar method of construction.

The method is based upon an analytic solution (Fotieva 1980) of elasticity theory flat contact problem for a medium simulating the rock massif, weakened a hole of an arbitrary shape (with a single axis symmetry), supported by a ring of another material simulating the lining of a tunnel having an initial stressed state with the inclined main axis to the vertical and the horizontal. The designed scheme is given in Fig. 1.

The problem was set as an inverse one and it was to evaluate such values λ, α, α^*, where the designed epure of the normal con-

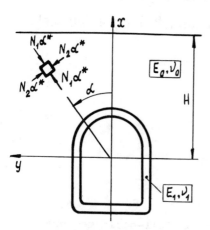

Fig. 1 The designed scheme

tact stresses σ_ρ best of all approach the measured one σ_ρ^* meaning the least quadratic deviation.

The method described has been programmed for the computer.

Results of processing the data of full-scale measurements received at constructing lining of one of the tunnels in stages is given below (the full-scale data were given by V.E.Nazarbegov).

The designs have been conducted for three sections along the tunnel length where measurement of normal rock pressure upon the lining in process of four stages of working sections opening were fulfilled.

The rocks around the tunnel for sections 1 and 2 had the following characteristics: the modulus of deformation E_0 = 18000 MPa, The Poisson ratio ν_0 = 0.23; for section 3: E_0 = 12000 MPa, ν_0 =0.23; while the characteristics of the lining material: E_1 = 30000MPa, ν_1 = 0.22.

On the works embracing the 1-st stage of the section opening a strengthening cementation of rocks has been conducted.

Designs have been conducted in the following order. First from every section at every stage processing the results of full-scale measurements of pressure upon a lining (a shute of concrete enduring a small stress was additionally introduced into the designed scheme which had an insignificant effect upon the results of solving the inverse problem and upon forces in

the upper part of the lining) was fulfilled and the designed characteristics of the stress initial field in massif: vertical $\sigma_x^{(0)}$, the horizontal $\sigma_y^{(0)}$ and the tangential $\tau_{xy}^{(0)}$ ones, the main stresses in an intact massif N_1 and N_2, their relation $\lambda = N_2/N_1$ and the α angle of the main axes inclination to the vertical and the horizontal have been evaluated. Then with the characteristics received being applied as the main basic data design of forces: bending moments M and longitudinal forces N in the lining sections have been conducted.At the same time the correcting multiplier α^* has been evaluated taking into account the conditions of initial vertical stresses being equal to the γH value (γ is the specific weight of the rocks, H is the working embedding depth), which was taken for four stages of opening, the sections being equal correspond to γH = 2.11 ; 2.16 ; 2.28; 2.28 MPa. Results of the processing data of full-scale measurements are given in Table 1.

Computation epures of normal contact stresses σ_ρ, forces M and N in section 1 of the lining designed at the basic of the Table 1 data are given for the 1-st and 2-nd stages of section opening in fig. 2,a,b; for the 3-rd and the 4-th ones in fig.3,a,b. Epures of the measured contact normal stresses are shown in those figures by dotted lines.

Similar epures have been received for cross-sections of the 2-nd and the 3-rd tunnels cross-sections.

As it follows from the results of processing the data of full--scale measurements of stress upon the lining in two cross-sections situated in similar mining-geological conditions (section 1 and 2) the designed characteristics of the initial stress field in the rock massif happen to be different for separate stages of the section opening. If initial stress field determined by measurement results performed after the 1-st stage of section opening is characterised by the fact that the greater main stress N_1 is nearer the horizontal one by its direction (α = 93°

Table 1. Results of elaborating full-scale measurements

Section number	Stage number	$\sigma_x^{(0)}$	$\sigma_y^{(0)}$	$\tau_{xy}^{(0)}$	N_1	N_2	λ	α	α^*
			MPa						
	1	0.12	0.22	−0.0047	0.22	0.12	0.55	93°	0.058
	2	0.29	0.32	0.021	0.33	0.28	0.85	63°	0.136
1	3	0.45	0.29	0.012	0.45	0.29	0.65	4°	0.199
	4	0.36	0.23	0.017	0.36	0.23	0.63	7°	0.158
	1	0.98	0.22	−0.018	0.22	0.096	0.43	98°	0.046
	2	0.31	0.39	−0.004	0.39	0.31	0.79	93°	0.142
2	3	0.42	0.44	−0.014	0.45	0.42	0.93	121°	0.186
	4	0.40	0.34	−0.011	0.41	0.33	0.83	171°	0.177
	1	0.055	0.20	0.002	0.20	0.055	0.28	89°	0.026
	2	0.21	0.31	0.044	0.32	0.19	0.59	69°	0.096
3	3	0.37	0.28	0.039	0.38	0.26	0.69	21°	0.090

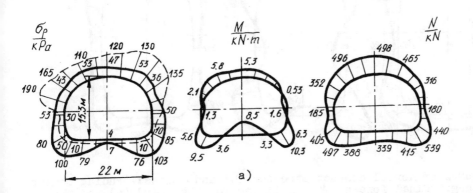

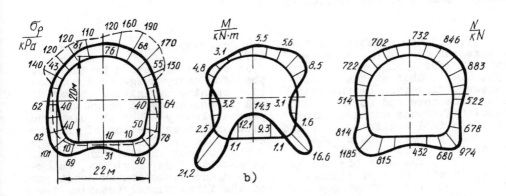

Fig. 2 Epures of normal contact stresses σ_ρ and forces: bending moments M and longitudinal forces N in lining for first (a) and the second (b) stages of the section opening

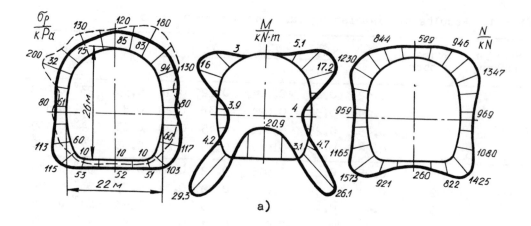

a)

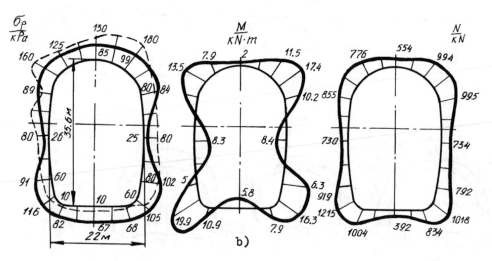

b)

Fig.3 Epures of normal contact stresses σ_p and forces: bending moments
M and longitudinal forces N in lining for the third (a) and the
fourth (b) stages of the section opening

and 98°), and the relation of the main stresses $\lambda = N_2 / N_1$ is substantially less than a unit (λ = = 0.55; 0.43), than at the following stages, direction of the main stress tends more and more to the vertical (after the working being opened to a full section $\alpha' = 7°$ and 171°), and the deviation of the main stresses approach one (λ = = 0.63; 0.83). The same picture is observed in section 3 of lining. In opening section 3 by stages the

α incline angle to the vertical changes from 89° (at first stage) till 13° upon the whole section being opened), while the relation

λ correspondingly from 0.28 to 0.73.

It can be explained by the fact that for every stage to come section opening by initial stress field is the stressed stage which appeared already after the distribution and concentration of the stresses at the previous working stage.

That is how in designing a lining of another form, rigidity, thickness or construction in conditions of the first tunnel (section 1 and 2) it is worthwhile to take the technique foreseeing opening of the face at once to a full section, to take into account initial vertical

116

stresses (the correcting multiplier α^* is to be taken into account) $\sigma_x^{(0)}$ = 0.12 MPa, while the lateral stress coefficient of rocks in an intact $\bar{\lambda} = \sigma_y^{(0)}/\sigma_x^{(0)}$ = 2.25 where more and more higher forces are received. If the section opening and the lining construction are assumed to go by stages, then for a final design of the whole lining one must assume $\sigma_x^{(0)}$ = 0.4 MPa $\bar{\lambda}$ = 0.83. In designing tunnel lining erected upon a full section in conditions similar to that of the second section (weaker rocks) one must assume for designing $\sigma_x^{(0)}$ = 0.055; $\bar{\lambda}_{(0)}$ = 3.6; at working done by stages σ_x = 0.31 , λ =0.73.

In general it appears that in rocks, surrounding the tunnel section under investigation there is an initial stress field of a tectonic origin where horizontal stresses exceed the vertical ones. It must be taken into account in designing linings in tunnels constructed in similar geotechnical conditions.

In fig. 3,a,b epures of normal tangential stresses $\sigma_{\theta int}$ upon an internal outline of the cross-section and forces: M and N arising in the lining of the first section, upon which measurements were made in erecting the tunnel upon a full section, are shown by dash lines. For comparison epures of stresses and forces appearing

at section opening by stages are shown by dotted lines.

As seen in fig. 3 at the working being opened at a full section normal tangential stresses $\sigma_{\theta int}$ in all points with the exception of the ones situated in the arched and the chute parts of the lining happen to be less, than at the opening by stages, tension stresses not exceeding 0.2 MPa appear.

The designs fulfilled allows us to come to a conclusion that for a lining of the shape investigated in geotechnical conditions from the point of view of the stressed conditions of the constructions an advancing of the working with the full section opening is more preferable.

CONCLUSIONS

The method given allows the volume of information to be widened, the one received from full-scale measurements, and the possibility of its being applied in practice. Designed characteristics of the massif determined as a result of the inverse problem being solved not depending upon parameters of the construction itself, they may serve as the basic data for calculating and designing linings of another shape of cross-section, thickness and rigidity in similar geotechnical conditions at similar construction

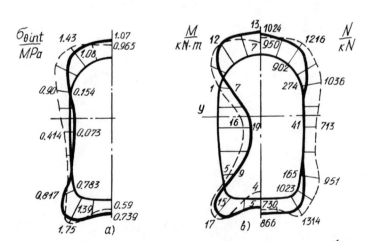

Fig.4 Epures of normal tangential stresses $\sigma_{\theta int}$ (a), bending moments M and longitudinal forces N (b) upon an internal outline of the lining cross-section (dashed lines in passing a full section; dotted lines in constructing by stages)

methods. Such a path allows the design results to be brought substantially nearer to the actual picture of the lining stressed state, and, therefore, leads to an increase in safety.

REFERENCES

Fotieva, N.N. 1980. Design of the underground structure support in seismic active regions. Moscow: Nedra.

Fotieva, N.N. & N.S.Bulychev 1980. An indirect method of determining stresses in rock massif on the basis of measuring the pressure upon the support of mining workings. Physical-technical problems of the natural resources being extracted. - N°5 : 26-29.

Fotieva N.N. 1981. Experimental analytical method of determination of massif estimated performance to design the underground structures support. Proc. of the 7-th Plenary Scientific session of the International Bureau of Rock Mechanics (World mining congress). Rotterdam: Balkema.

Fotieva N.N. & N.S.Bulychev 1982. Indirect method of measuring initial stresses in solid rocks for construction of tunnels and mines Proc. of the Conference of Construction Practices and Instrumentation in Geotechnical Engineering. Surat (India).

Effects of Geomechanics on Mine Design, Kidybiński & Dubiński (eds) © 1992 Balkema, Rotterdam. ISBN 90 5410 040 0

Interpretation of the results of full-scale measurements in the vertical mines shafts

N.S. Bulychev & I.I. Savin
Tula Polytechnical Institute, Russia

ABSTRACT: Design of multi-layer lining of round cross-section vertical mines' shafts by the results of stress measurements (normal and tangential) and deformations upon an arbitrary contact of a monolayer lining or upon a contact of the lining with the rock body undergoes investigation in work presented. A specific feature of the design is the possibility of fully reestablishing stresses in lining by the results of measuring some arbitrary kind of stresses or deformations.

The design is made applying the elastic model of contact interaction of the lining with the rock body. The rock body appears a linearly deformed medium. Deep embedding working $H \gg R$ (H – the depth of bedding, R – the radius of working) is under investigation. The inverse problem undergoes its solution: stress initial field specifications are estimated by the normal radial or tangential stress known, deformation or displacements (coefficient of non-uniformness and direction of the main maximum initial stresses $\tilde{\alpha}$, the coefficient α^* which takes into account the lining erection lagging behind rock exposure).

For determining unknown quantities we form taking the results of measurements into consideration equation system redistribution in the form of

$$a_{j1} P_o^{eq} + a_{j2} P_2^{eq} \cos 2(\Theta_j - \tilde{\alpha}) = P_j^* \quad (j=1,2,...,N)(1)$$

where P_o^{eq} and P_2^{eq} are the equivalent stresses constituents having called a stress-deformed condition of the lining; a_{j1} and a_{j2} are the coefficients depending upon the measurements results; $\tilde{\alpha}$ is the angle characterising the direction of the main maximum initial stresses; P_j^* is the quantitative characteristic of the size being measured; Θ_j is the angled coordinate of the size being measured.

Taking equation (1) system into consideration we evaluate angle $\tilde{\alpha}$ and the coefficients

$$\xi = \frac{P_o^{eq} \mathcal{H}_o - 2 P_2^{eq}}{P_o^{eq} \mathcal{H}_o + 2 P_2^{eq}} ; \qquad (2)$$

$$\alpha^* = \frac{P_o^{eq}(\mathcal{H}_o + 1)}{\lambda \gamma H (1 + \xi)} ,$$

where $\mathcal{H}_o$ is the coefficient of the type of rock body stresses state

$$\mathcal{H}_o = 3 - 4 \nu_o$$

ν_o is the rock Poisson coefficient; γ is the weight of the rock in a volume unit, MH/m^3; H is the depth the working is sunk at, m.

Depending upon the characteristics of the size being measured coefficients a_{j1} and a_{j2} are evaluated as follows:
1. In case if there N measurements of normal radial stresses upon an external outline of an arbitrary i-th layer the coefficients stated above are estimated by the formulae

$$a_{j1} = W_{0(i)} ;$$

$$a_{j2} = W_{2(i)} ,$$

where $W_{0(n)} = k_{0(n)}$; $W_{2(n)} = k_{11(n)}$;

$$z_{2(n)} = k_{21(n)};$$

$$\{W_i\} = [K_i] \times \{W_{i+1}\};$$

$$\{W_i\} = \begin{Bmatrix} W_{0(i)} \\ W_{2(i)} \\ z_{2(i)} \end{Bmatrix}; \qquad [K_i] = \begin{bmatrix} k_{0(i)} & 0 & 0 \\ 0 & k_{11(i)} & k_{12(i)} \\ 0 & k_{21(i)} & k_{22(i)} \end{bmatrix}$$

Here $k_{0(i)}, k_{11(i)}, k_{12(i)}, k_{21(i)}, k_{22(i)}$ are the coefficients of external loads transfer /1/; θ_i is the angle of coordinate of the point of the i-th measurement ($j = 1, 2, \ldots, N$); P_i^* is the value of measured normal radial stresses upon an external outline of the i-th layer.

Shaft lining design of the "Centralnaya" coal mine elaborated by the results of radial stresses measurements upon the contact of lining with the rock body. The results of the design are given in fig.1.

2. If there are N measurements of normal tangential stresses $\sigma_{\theta(i)}^j$ upon the internal outline of the arbitrary i-th layer, the unknown coefficients α_{j1}, α_{j2} are estimated by the formulae:

$$\alpha_{j1} = W_{0(i+1)} m_1 - W_{0(i)} m_2;$$

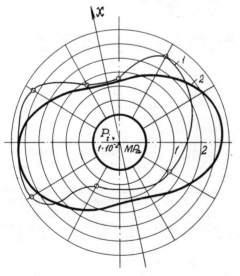

Fig.1
1 is the epure of measured stresses
2 is the epure of designed stresses

$$\alpha_{j2} = -(W_{2(i+1)} n_1 - z_{2(i+1)} n_2 - W_{2(i)} n_3 + z_{2(i)} n_4),$$

where coefficients m_1, m_2, n_1, n_2, n_3, n_4 are estimated for i-th layer coming from the geometric specifications of the layer in question /1/.

Epures of measured and designed normal tangential stresses upon the cage shaft concrete lining internal outline of the "Centralnaya" coalmine of the Donskoy GOK are given in fig.2.

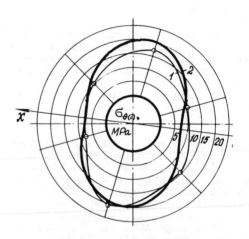

Fig.2
1 is the epure of measured stresses
2 is the epure of designed stresses

3. If radial displacements (convergences) $u_{(i)}^j$ of the internal outline points of the arbitrary i-th layer of a multilayer lining have been measured, the coefficients α_{j1}, α_{j2} are estimated by the formulae:

$$\alpha_{j1} = \frac{R_i}{4 G_i (c_i^2 - 1)} (W_{0(i+1)} d_{1(i)} - W_{0(i)} d_{2(i)});$$

$$\alpha_{j2} = \frac{R_i}{12 G_i D_i} [W_{2(i+1)}(\alpha_{1(i)} + 3\alpha_{1(i)}') - z_{2(i+1)}(\alpha_{2(i)} + 3\alpha_{2(i)}') - W_{2(i)}(\alpha_{3(i)} + 3\alpha_{3(i)}') + z_{2(i)}(\alpha_{4(i)} + 3\alpha_{4(i)}')],$$

where R_i is the internal radius of the lining layer being designed m; G_i is the shear modulus of the lining layer material being designed, MPa; coefficients D_i, c_i, $\alpha_{1(i)}, \alpha_{2(i)}, \alpha_{3(i)}, \alpha_{4(i)}, \alpha_{1(i)}', \alpha_{2(i)}', \alpha_{3(i)}', \alpha_{4(i)}', d_{1(i)}, d_{2(i)}$ are estimated for lining layer being designed in accordance with /1/.

4. If measurements of tangential deformations $\varepsilon_{\theta(i)}$ upon the internal outline arbitrary i -th layer are conducted, the coefficients a_{j_1}, a_{j_2} estimated by the formulae:

$$a_{j_1} = \frac{1}{2G_i}\left\{W_{o(i+1)}m_1(1-\nu_i) - W_{o(i)}[m_2(1-\nu_i)+\nu_i]\right\};$$

$$a_{j_2} = \frac{1}{2G_i}\left\{(W_{2(i+1)}n_1 - z_{2(i+1)}n_2)(1-\nu_i) -\right.$$

$$\left. - W_{2(i)}[n_3(1-\nu_i)-\nu_i] + z_{2(i)}n_4(1-\nu_i)\right\},$$

where ν_i is the Poisson coefficient of material of the lining layer in question.

Epures of the measured and designed relative and tangential deformations upon the internal outline of the back of the "Centralnaya" coal-mine of the Donskoy GOK shaft cast-iron tubing lining.

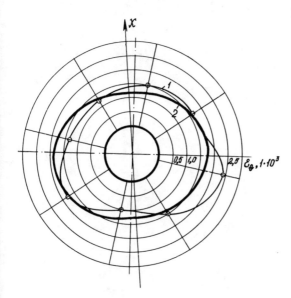

Fig.3
1 is the epure of measured deformations; 2 is the epure of designed stresses

do not depend on the geometric and deformation specifications of the lining in which the measurements were conducted and as a result of independence mentioned can be used for designing other construction of lining, operating in **similar** conditions.

REFERENCES

Bulychev, N.S. Mechanics of underground construction. Moscow, "Nedra", 1982. 270p.
Bulychev, N.S. & Savin, I.I. Technique of field measurement interpretation in erecting underground openings of round section. Proc. of 2-nd Int. Symp. on Field Measurements in Geomech. In S.Sakurai (ed) (1988): 1225–1230. Rotterdam, Balkema.
Fotieva, N.N. & Bulychev, N.S.1979, Using data of full-scale measurement in lining design for underground structures. Proc., Fourth congr. of the Int.Soc., for Rock Mech.: Montreus, Switzeland, Sept. 02.08, vol.1: 387–392.

CONCLUSIONS

The specific feature of the design technique given is that specifications of the initial stresses state of the body ξ, α^* and $\tilde{\alpha}$ estimated as a result of the design

3 Design considerations and mining technology improvements

Effects of Geomechanics on Mine Design, Kidybiński & Dubiński (eds) © 1992 Balkema, Rotterdam. ISBN 90 5410 040 0

Mining design of a vein orebody exploitation using two cemented fill methods

O. Del Greco, A. M. Ferrero, G. P. Giani & D. Peila
Dipartimento di Georisorse e Territorio, Politecnico di Torino, Italy

ABSTRACT

The paper refers to the mining design of a mixed sulphide vein orebody. The main aspects of the mine design involve geomechanical and hydrogeological problems.
The studied methods used the cemented filling technique.
The aim of the paper is to describe a design method based on the calibration of the desig geomechanic model according to the results of the experimental in situ and laboratory measurements.
In situ control measurements were carried out to monitor the rock mass, rock supports and cemented fill static behaviour during the mining exploitation in the experimental panels.
Laboratory determinations were carried out to asses the strength and deformation features of the intact rock, rock discontinuities, supports and cemented fill.
The geomechanical model involved, the numerical simulations of the two different mining stopes during the experimental exploitation activity.
The numerical simulation were carried out by using the finite elements method in two and three-dimensional conditions and in the elasto-plastic field to analise the statical behaviour of the underground rock excavation.

1. INTRODUCTION

The aim of the research has been to develop the study of the stability conditions connected to the exploitation design of the lead and zinc Molaoi mine located in the Lakonia province of Peloponnesos (Greece).

The mining design was sponsored by a European Economic Comunity research contract and also involved two different mining methods applied in two experimental pannels. In the research process were involved the Aegean Metallurgical Industries SA (Greece), Institute of Geology and Mineral Exploitation (Greece), Mining Sector - National Technical University of Athens (Greece) and the Dipartimento di Georisorse e Territorio of the Politecnico di Torino (Italy).

Becouse of the scarce quality of the rock mass hosting the orebody, the hydrogeological conditions of the area and the geometry of the mineralized mass, the choice of the exploitations method has been devoted to the cemented fill method application.

To controll and define the rock mass characteristics: laboratory tests for the determination of the strength and deformation features of the rock material, of the rock mass and of the wooden support was carried toghether with a detailed structural surveying.

A field measurement instrumentation for the control of the mechanical behaviour of the rock mass during the exploitation, has been studied and installed. Finally the numerical simulation of the exploitation activity in the two experimental pannels of the Molaoi mine were carried out by means of three Finite Element Method (FEM) models.

The numerical models set up for this simulation purpose was calibrated with the results of the experimental measurements and the engineering parameters for the future mining design was then gathered.

2. GEOLOGICAL CONDITIONS OF THE AREA

The Molaoi mine deposit of sphalerite followed by pyrite and galena, belongs to the stratiform volcanogenic massive sulphite deposits that where formed in a submarine environment. The volcanic host rocks consist of altered tuffs, tuffites and other pyroclastic lavas as well as massive aphanitic and porphyritic lavas which have an andesitic to basalt-andesitic composition and show a calc-alkaline character. Mineralization consists of three stratiform type orebody that strike N-S and dip 55-60 to the East with thickness varying between few centimeters to about 10 m.

On the basis of several studies four different rock zones have been recognized in the rock mass (figure 1) :
  - A discontinuous rock zone formed by volcanic rock extensively fractured by shear faults (material A);
  - a transitional zone formed by material of type A and mylonite (material B);
  - mylonite zone (material C) its thickness ranges from approximatively 0.5 and 1 m and is in direct contact with the mineralization;
  - a vein type polymetallic mineralized zone (material D).

All the different material involved need to be studied in order to determine their mechanics behaviour.

An other important rock mass feature is the hydrogeological condition : boreholes drilled during mineral exploration revealed

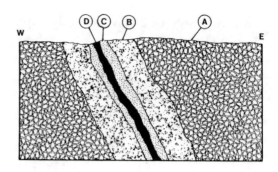

Fig. 1 Schematic geological rapresentation of the mineralized formation in Molaoi area

initial artesian or sub artesian water. Geological condition in combination with the permeability and area tectonic create a steady water level in the mine area. Places with water storage capacities are faults and mineralized zones.

Water table position has been studied in order to take in account dewatering phenomena during the mining activities.

3. EXPERIMENTED MINING METHODS

The experimented mining methods where:
a) top slicing with cemented fill
b) long hole drilling with cemented backfilling

The ore access is obtained with a ramp 400m long (12% dip) starting at the +165 level (surface)

The a) method has been applied between the drifts +135 and +129 in the north section of the mine (fig. 2).

The entire section was filled before starting the excavation in the lower slice. During the filling operation was not possible to recover the supports.

This method did not present any stability problem.

The b) method has been applied in the south section of the mine between the levels +135 and +129 excavated in the direction of the orebody (fig. 2).

The work was carried on driving the excavation from the lower level and then filling the hole before going on with the excavation.

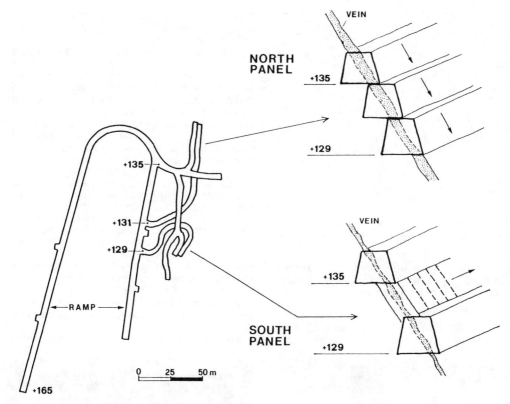

Fig. 2 Experimented mining methods.

This method had from the beginning problems of stability of the roof of the drifts.

4. EXPERIMENTED WORKS

The structural analysis of the rock mass was carried out on the tunnels sites studing the discontinuities conditions and the rock mass was classified according to the Bieniawski criterium. The results obtained are very varables (fig 3) and the rock mass can be classified as a medium-poor quality one.

From these data the rock mass elasticity modulus was defined using the Serafim & Pereira (1983) relation.

$$Em = 10^{(RMR-10)/40} \ [MPa]$$

The laboratory analysis involved both the wood of the supports (uniaxial compression and flexural tests) and the rock material (uniaxial compressive tests, direct shear tests). The uniaxial strength of the rock was tested using a rigid MTS press.

For the tested woods the obtained uniaxial compressive strength

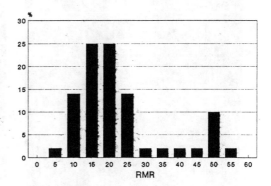

Fig. 3 Rock mass rating distribution in the Molaoi mine

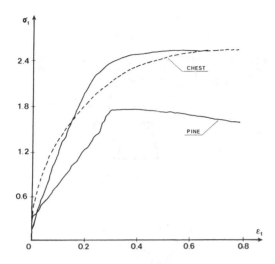

Fig. 4 Stress-strain curves obtained by the compressive tests of the wooden specimens

values (σ_c) and elastic moduli are collected in table 1 and the

figures 4 and 5 show the stress-strain curves obtained and the laboratory device.

The free span of the flexural tests of wooden beams were fixed in 200 cm, according to the geometry of the support frames used in the Molaoi mine. The load scheme is indicate in fig. 6.

The control measurement carried on in the mine have been located according to the schemes of fig.7. The measurements have been:

- convergence stations in the panel adits , both on wall tunnels and on the supporting steel arches: the station was composed of two or three pins bolted in the rock or on the supports as soon as possible.

The distance between the pins was regularly measured at time intervals, in order to determine the convergence of the supports and of the walls of the tunnels.

The instrument employed is a mechanical convergence measuring device.

The measurements sections were

Fig. 5 Laboratory device of the chest specimens before and after the collapse

Table 1. Results of the laboratory tests on wood

Wood	Compressive test			Flexural test	
				bearing beam	fixed-end beam
	σc	Et	Es	σ	σ
	(MPa)	(MPa)	(MPa)	(MPa)	(MPa)
Pine	25.58	5541	15096	46.03	27.29
Pine	17.83	6809	10121	79.28	47.00
Chest	25.50	8079	10834	87.31	51.77
Chest				80.20	47.55

Fig. 6 Flexurale test in progress

chosen where the greatest values of displacements were expected and the influence of the influence of the exploitation works were supposed to be more significant.
- dynamometric measurements on the supports:
the device is a vibrating wire dinamometers and was installed on the steel supports to measure the loads;
- measurements deformation of the steel supports:
the deformation of the intrados of the steel arches were made up using a segment deformation measurement and a radial ond. The radial deformometer measures indirectly the change in curvature of a straight or curved beam at a given point.

- extensometer measurements:
long hole extensometers were installed in two section of the mine to measure the different movements inside the rock mass and to define the extension of the plastic area.
Three borehole extensometer with three points anchor each were installed in a single section In the south sector of the mine.
In the north part two BHE of two point each were installed in the hanging wall and in the footwall of the orebody.
- inclinometric measurements:
this device was installed from the surface in order to gather horizontal displacements of the rock above the north panel. During

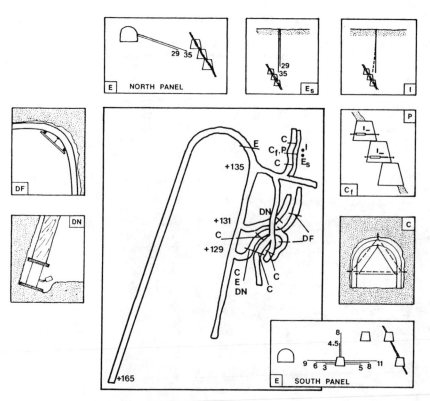

Fig. 6 Flexurale test in progress

the escavation of the +135 level no displacements were recorded;
 - hydraulic pressiometric cells: these cells were installed to determine the stresses in the fill material in the middle of the north pannel slices (upper and middle ones);
 - displacement measuraments in the fill material:

this measurements were carried on by means of electrical extensometers installed in the cemented fill of the north pannel near the pressiometric cells.
 That device was expecially build up by the Dipartimento di Georisorse e Territorio of the Politecnico di Torino.

Table 2 - Mechanical characteristics of the rock mass and supports

Rock mass	E (MPa)	ν	α	K (MPa)
MODEL 1	350	0.31	0.181	0.32
MODEL 2 a	350	0.375	0.23	0.117
MODEL 2 b				
material 1	112	0.375	0.29	0.00116
material 2	200	0.375	0.23	0.0866
MODEL 2c				
material 1	112	0.375	0.23	0.00115
material 2	200	0.375	0.23	0.0866
MODEL 3				
vein	2370	0.23	0.181	0.00001
rock	2370	0.23	0.233	0.00001

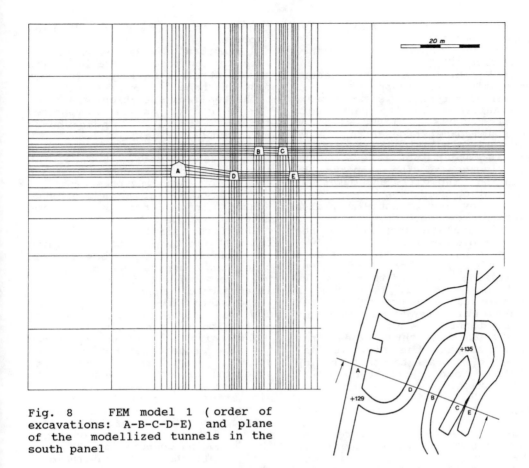

Fig. 8 FEM model 1 (order of excavations: A-B-C-D-E) and plane of the modellized tunnels in the south panel

5. NUMERICAL MODELS

Three different numerical models were set up in order to analize the stress-strain behaviour of the mining structures.

One of the models concerns the preparatory works and the exploration activity of the experimental south panel (MODEL 1, fig . 8)

The other two (MODEL 2 a,b,c; fig. 9 and MODEL 3; fig. 10) simulate different phases of the exploitation activity in the northern stope respectively in a 3 dimensional and in a 2 dimensional field. Deformation and strength parameters used in order to simulate the behaviour of the rock mass in the 3 models are reported in table 2.

The geotechnical behaviour assumes that the rock mass is elastic ideally plastic with plasticity law defined by Drucker and Prager plasticity surface. The plasticity law is an associated type. The behaviour of the excavation supoorts is assumed linear elastic. The model 1 includes the rock mass from the countryside surface down to a 88 m depth. The model is 120 m large and is made up of 1008 quadrilateral elements and 16 beam elements. Quadrilateral elements represent the rock mass, while beam elements represent the supports.

The numerical simulation of the mining activity carried out in the south panel, for the preparatory works, involves the following FEM analysis:
- gravitational analysis;
- ramp excavation (A);
- installation of the first ramp supoorts;
- explorative tunnel excavation(B);

131

- installation of explorative tunnel supports;
- second explorative tunnel excavation (C);
- installation of the second explorative tunnel supports;
- panel adit tunnel excavation (D);
- installation of the panel adit tunnel supports; - first slice of the south panel excavation. Two numerical simulations are carried out in order to assess the effectiveness of the supports, by comparing two different numerical exploitation sequences :
a) the above quoted activity phases in absence of supports;
b) the complete sequence of the phases above described.
It has to be pointed out, that as a 2-D model has been used, any stabilizing effect of the advancing front is neglected. The model is not able to reproduce other three dimensional effects due to cross cutting tunnels.

The MODEL 2 is three dimensional and represents the first tunnel excavated in order to start the top slicing and cemented fill exploitation.

The model is 60 m large 125 m deep from the surface and 45 m thick and is made up of 400 elements. The 3-D model represents:
- the mining tunnel at the upper level;
- the bolts and the wooden supports;
- the rock mass surrounding the tunnel up to the countryside surface.

A numerical simulation is set up with the purpose of examining the stress-strain behaviour of the rock mass in three different situations.

The analized situations are :
- excavation of the unsupported tunnel in a homogeneous material (MODEL 2a);
- excavation of the tunnel in a homogeneous material by applying the correspondent supports for each excavation advancing phase;
- excavation of an unsupported tunnel in a non homogeneous material (MODEL 2 b,c).

The first simulation is carried out with the purpose of determining the behaviour of the rock mass in order to compare the exacavation wall convergences in presence and in absence of supports.

The second simulation reproduces the effective sequence of the excavation and supporting phases.

The third simulation reproduces the same sequence of phases as the second, but the rock mass schematization is presumed to be non homogeneous. In the third simulation a 18 m thick weaker zone surrounding the tunnel was assumed.

This zone has a vertical narrow shape. The simulations concernes the following computation steps:
- application of gravitational state of stress ;
- excavation of 1 m tunnel length;

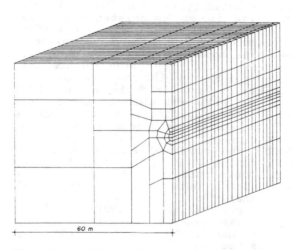
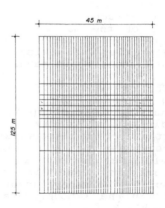

Fig. 9 - FEM model 2:north panel

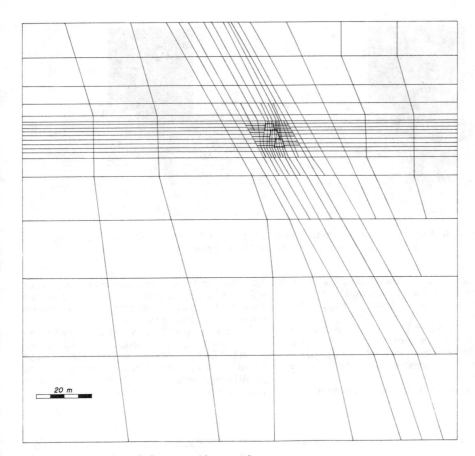

Fig. 10 - FEM model 3:north panel

- excavation of 1 m long tunnel;
- repetition of excavation and supporting phases up to the end of the tunnel costruction.

The third model (MODEL 3) is 2D and is carried out in order to simulate the numerical behaviour of the north mining stope .

The model includes the rock mass from the surface down to a 140 m depth and it is 160 m large.

The model is made up of 446 quadrilatal elements and 9 truss elements.

The simulation reproduces the effective sequence of the following excavation and supporting operations:
- excavation of the exploitation tunnel,
- installation of the supports,
- placinng of the cemented fill. These operations are repeated for the three tunnels in the descending order. The drainage due to mining works is considered in the simulation. An increase of effective pressures corresponds to a decrease of hydraulic pressures.

In the numerical modelling an increase of gravitational forces was applied to simulate the progressof the drainage due to the exploitation activity.

In particular, after every excavation phase, body forces are applied in connection with the design water table decrease.

6. COMPARISON BETWEEN MODELLING AND MEASURING RESULTS

6.1 South Panel

The most significant results, obtained with model 1, is the plasticity zones around the stopes. Figure 11/a represents the plasticized areas without supports

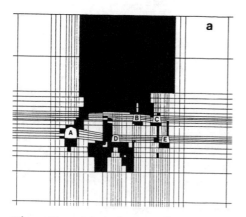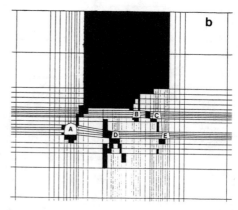

Fig. 11 - Plasticity zones around the tunnels of the south panel (model 1). Case a) without supports, case b) with supports

while fig. 11/b represents the plasticized areas obtained in the analysis with the supports modelling.

The comparison between plasticity areas computed in the presence and absence of supports allowed one to make the following considerations:

- the plasticity areas determined in the case of unsupported tunnels were always larger, especially in the zone surrounding the woodden supported tunnels;

- the plasticity areas reached the countryside surface in both simulations. This situation could lead to serious subsidence problems, even though they have not yet occurred. The subsidence

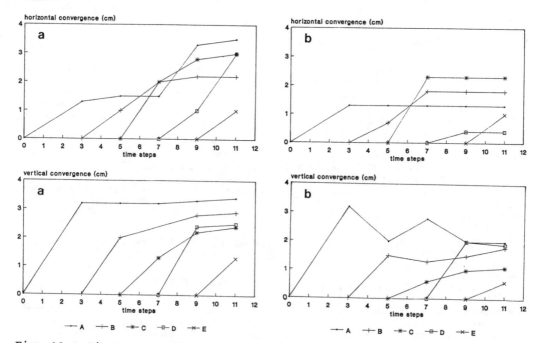

Fig. 12 - Diagrams of the computed displacements of the tunnels of the south panel (model 1).Case a) without supports, case b) with supports

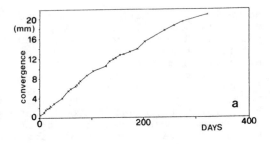

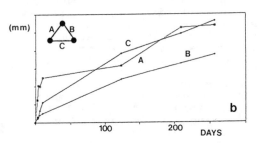

Fig. 13 Convergence measured values
in the +129m tunnel (south panel)
a) convergence of the rock
b) convergence of the steel support

results:the excavation of the
explorative tunnels (B and D)
induced an increment of the
horizontal convergence displacement
up to 3.5 cm and an increment of
the vertical convergence
displacement up to 3.35 cm,in the
absence of supports in the tunnel
A.

The analysis,in the presence of
supports, did not provoke a
remarkable reduction of the
horizontal convergences, while the
vertical convergences were reduced
to 2 cm values.

The computed bending moments and
normal stresses did not
provoke instability phenomena in
the wooden supports (the elastic
stresses computed in the supports
were always compatible with the
strength features of the wood).

These results are comparable with
the load measured values obtained
on steel arches in a section near
the simulated one (the maximum
load value is about 3000 N) (Fig.
14).

problems shown by the numerical
simulation were due to a too
conservative strength parameter
choice; such an extended plasticity
area could cause a collapse of the
mining panel. The exploitation of
the south panel was in fact
interrupted for technical
difficulties in executing the
foreseen mining programs. For this
reason the prosecution of the
numerical simulation including the
vein exploitation and the cemented
fill application was not carried
out.

Comparing the computed results
(Fig. 12) and the measured values
(Fig.13) in the mine,the following
cosiderations can be made:
- the walls of the mine ramp (A)
had a horizontal convergence
displacement of 1.3 cm and a
vertical convergence displacement
of 3.15 cm at the end of the
excavation and before the
support installation.

These computed values reach
stabilization while the measured
ones increase in time; the computed
values are constantly lower than
the measured ones.

Some considerations can be made
by means of the modelling

6.2 North panel

For the 3-D model, the analysis of
the experimental and numerical
results allows one to make the
following considerations:
- The homogeneous medium
assumption led one to overvalue
effective stiffness of the rock
mass. Moreover the assumed
strength parameters led to
determine a limited plastic
zone surrounding the tunnel and
consequently the stability of
the rock tunnel was also reached
for the condition of the

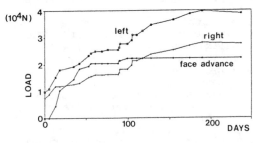

Fig. 14 - Diagrams of measured
loads on the steel arches of the
+131 tunnel (south panel)

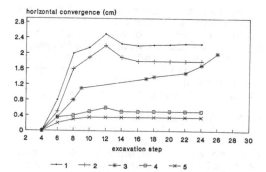

horizontal convergence (cm)

excavation step

—+— 1   —+— 2   —*— 3   —□— 4   —×— 5

Fig. 15 - Horizontal measured convergence compared with the calculated with the FEM model
Key:
1) homogeneous unsupports
2) homogeneous with supports
3) measured convergences
4) non homogeneous unsupported (set of parameters 2)
5) non homogeneous unsupported (set of parameters 1)

unsupported tunnel.
- The non homogeneous rock mass material assumption led to a more realistic situation. However, in this case, the stiffness of the equivalent medium was lower than the effective rock mass stiffness. Consequently, for the shorter distance between the excavation face and the measurement station, the computed convergences were greater than the measured ones. The developed situation for non homogeneous rock mass, did not take into account the support application which can reduce the computed measurements till a value comparable to the measured ones. The numerical simulation of the advancing excavation led to the stability of the wall convergences (Fig. 15).These results were in contrast with the experimental results which show a continuous increase of the convergence even for wide distances between the excavation face and the measurement station.
The contrasts between experimental and computed results proved that a more realistic stress-strain law should have been assumed in order to obtain a better

agreement with measured convergences. In particular, the fact that the wall convergence did not reach stability even for a wide distance from the advancing front could indicate a swelling behaviour of the rock mass that is not modellized with our computer code.
An other important factor, that has already been considered in the model 3, was that the water drainage effect that induced a large variation in the effective stresses distribution around the tunnel area. This variation could determine an increment in the convergence displacements able to better fit the measured convergences.
According to the numerical simulations of the model 3 the following considerations can be made:
- the plasticity areas had an elliptical shape around the excavated vein (Fig.16). The maximum ellipse axis was subvertical, while the minimum was sub-horizontal. The extension of the plasticized area did not reach the ground surface and remarkable mining subsidence did not occur in the model and was not measured with the installed instruments.
The plasticized zone which had been numerically determined seemed to be in agreement with the extensimeter measurement interpretation.
- the walls of the upper tunnel were subjected to a horizontal convergence displacement of 0.5 cm after the excavation and supporting procedures (Fig. 17). These values

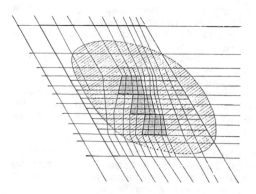

Fig. 16 - Plasticity zones around the north panel (model 3)

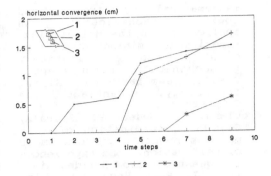

horizontal convergence (cm)

— 1   —+— 2   —*— 3

Fig. 17 - Horizontal convergences of the north panel (model 3)
Time step 1) excavation of 1
Time step 2) excavation of 2
Time step 3) excavation of 3

are lower than the measured values which after 300 days reach 1.5-2 cm and they do not show any stabilization (Fig. 18).

At the end of the panel exploitation the computed displacements increased up to 1.5 cm.

The computed increase of displacement of the upper tunnel

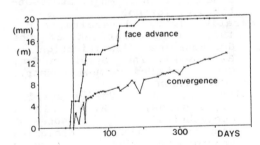

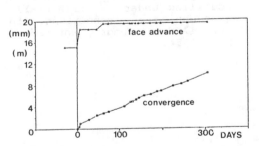

Fig. 18 - Measured convergences in the upper tunnel of the north panel

filled with the cemented fill due to the excavation of the lower level is of 0.9 cm while the measured values are 1.4 cm (Fig.19).

The walls of the second tunnel were subjected to a horizontal convergence displacement of 1.0 cm after the excavation and supporting procedures. This displacement increased up to 1.66 cm at the end of the panel exploitation. The walls of the third tunnel were subjected to a horizontal convergence displacement of 0.3 cm after the excavation and supporting procedures; the wall convergence displacements were still lower than the measured values, even though in a better agreement than the previous models. The numerical simulation of the drainage effect seemed to be determinant for the best fitt of the physical rock mass behaviour.

7. CONCLUSIONS

All the studies carried on, based on direct observations, monitoring measurements and numerical modelizations, focus that the choise of a filling exploitation method is the only possible one (because of the complex rock mass structure).

In particular, a cemented filling material has been chosen because of its strength and stiffness characteristic. The cement presence anable the filling to support the

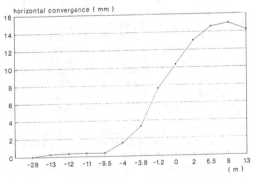

horizontal convergence (mm)

Fig. 19 -Measured horizontal convergences in the filled upper tunnel due to the excavation of the lower level

137

high pressure applied by the weak embedding rock formation.

The top-slicing method, in the north panel, gives better results, between the two experimented methods, because of the reduction of the mining works.

REFERENCES

Barton, N.R.,Lien,R.,and Lunde, J., 1974.Engineering classification of rock masses for the design of tunnel support. Rock Mech. Vol. 6, 189-239.

Bieniawski, Z.T., 1974. Geomechanic classification of rock masses and its application in tunnelling. 3rd Int. Cong. on Rock Mech., Denver, Vol.2, part A, 27-32.

Descoeudres, F.,1977. Mecanique des Roches, ISRF-EPFL Lausanne.

Donato, D., 1989. Studio di un pannello sperimentale per la coltivazione di un giacimento stratiforme raddrizzato sottile di Pb-Zn. Tesi di Laurea,Politecnico di Torino.

Hoek, E.,1990 .Estimating Mohr-Coulomb friction and cohesion values from the Hoek-Brown failure criterion. Int. J. Rock Mech.Min. Sci. & Geomech. Abstract, Vol. 27 227-229.

Hoek, E.and Brown, E.T., 1980. Underground excavation in rock. IMM, London.

Ferrero, A.M. , Giani, G.P., Peila, D. , Vigna, B., 1990. Analisi e studio di cantieri sperimentali per la progettazione della coltivazione della miniera di Molaoi. I Convegno Minerario Italo-Brasiliano, Cagliari.

Hoek, E. and Brown, E.T., 1980. Empirical strength criterion for rock masses. Int. J. Geotech. Engg Div. ASCE 106 1013-1035.

Hoek, E. and Brown E.T., 1988. The Hoek-Brown failure criterion 1988 update. Proc. 15th Can. Rock Mech. Symp., University of Toronto, 31-38.

Hoek, E., 1983 Rankine lecture, strength of jointed rock masses. Geotechnique Vol. 33, 187-223.

Kontopoulos, A., Katsifos, A., Vardakastanis,D.,1990. Optimization of the exploitation of thin vein polymetallic sulphide deposits through mathematical modelling and rock mechanics; an application to the Molaoi Mine. Experimental mining works, in-situ and laboratory geomechanical measurements. C.E.C. Seminar on Mining Technology, Santiago de Compostela.

Krouland, M., Soder, P., Aginalm, G.,1989.Determination of rock mass strength by rock mass classification. Some experiences and questions from Boliden mine. Int. J. Rock Mech. and Min. Sci.& Geomech. Abstr., Vol. 26, 115-123.

METBA, 1988-1990.Technical reports for the project:"Optimization of the exploitation of thin vein polymetallic sulphide deposits through mathematical modelling and rock mechanic; an application to th Molaoi mine. Unpubblished reports 1,2,3.

Olofsson, T., 1985. Mathematical modelling of jointed rock masses. Doctoral Thesis, Lulea University, Sweden.

Pelizza, S., Del Greco, O., Ferrero, A.M., Giani G.P., Manfroi, I., Moreau, P., Schmidt H,.1989. Sviluppo di un nuovo metodo di coltivazione di talco: studi e controlli. Int. Congr. Suolosottosuolo, Torino.

Serafim, J.L. and Pereira, J.P., 1983. Considerations of the Geomechanics classification of Bieniawski. Int. Symp. on Eng. Geol. and Underground Constr., LNEC,Lisbon.

Tzitziras, A., Vardakastanis, D.,Katsifos, A., 1990. Geotechnical measurement campain for the designing underground works through modelling under unfavorable geotechnical conditions. 6th Int. IAEG Congr., Rotterdam, 2549-2554.

Yield pillars in a deep potash mine

P.Squirrell
Cleveland Potash Limited, UK

ABSTRACT: This case study describes the evolution of the yield pillar mining technique used at Cleveland Potash Ltd's Boulby mine and proposes a hypothesis as to how the system works.

1 INTRODUCTION

Potash salts were first discovered in 1939 at Aislaby in North Yorkshire by the D'Arcy Exploration Company while test drilling for natural gas. The deposits were not further explored however until after 1945 but those found, around Whitby, were thought to be too deep for economic development. In 1965 exploration was reopened North-West of Whitby, around Staithes, where it was thought that the seam may be at a shallower depth. Drilling proved this to be so and found Potash at around 1100 metres below surface. In 1969 Cleveland Potash Limited commenced shaft sinking and the first Potash was produced in 1973, though it was not until 1977 that both shafts were commissioned.

Table I shows the ore hoisted each year since 1973 and it is clear that the planned build up to 2.5 million tonnes was slow. The main reasons for this were the technical problems of mining a weak rock at depth and the choice of mining equipment.

This case study is about the evolution and operation of the yield pillar mining technique used at Boulby, which enabled the difficulties to be overcome.

The history of Boulby Mine and a more detailed description of the geology have been described in several papers to which the interested reader is referred (see references).

2 GEOLOGY

Figure 1 shows the typical geological succession at Boulby Mine. The Potash seam dips from North to South and is 900 metres deep in the North and

Table I

YEAR	HOISTED ORE
1973	48,025
1974	210,250
1975	347,751
1976	567,587
1977	799,608
1978	945,260
1979	1,502,094
1980	1,836,880
1981	1,655,496
1982	1,419,160
1983	1,601,520
1984	1,789,200
1985	1,985,204
1986	2,115,780
1987	2,353,088
1988	2,507,251
1989	2,592,783
1990	2,608,765

1250 metres in the South. It is a series of gently dipping steps joined by overfolds which are associated with faulting in the sedimentary rocks above and below the evaporites.

The Potash ore is a mixture of Sylvite (20 - 60%), Halite (30 - 50%) and Insoluble materials such as Snale and Anhydrite (5 - 30%). The average Potash ore is 37.5% Sylvite, 50% Halite and 12.5% Insolubles but it is variable and can change quickly within the limits shown. This variation is evident through the seam with often a high Sylvite/low Shale ore at the base of the seam with the Sylvite decreasing and the Shale increasing towards the top of seam. In some areas of tne mine the low Shale ore is completely absent and on the base of seam the ore is low Sylvite/high Shale with the Sylvite decreasing Shale increasing towards the top of

Metres below surface

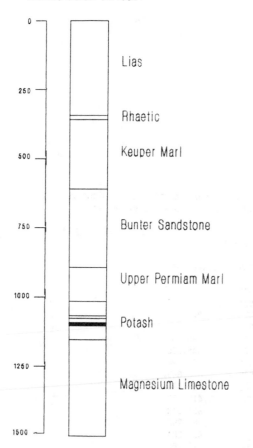

GEOLOGICAL SUCCESSION

Figure 1

seam. The roof material is usually weaker than the pillar material and the increase in Shale content means that there are often lenses of Shale in the roof with little or no strength in the interfaces between these and the Sylvite and Halite rocks.

The Middle Halite is about 95% NaCl with Shale (2 - 5%) and Anhydrite (0 - 2%) being the main impurities. Crystal size is also variable and this, the impurities and foliation affect the strength of the Halite.

3 ROCK PROPERTIES

A programme of rock testing, summarised in table 2 and 3 and Figures 2 and 3, was undertaken by Newcastle University. (Buzdar, S.A.R.K, (1967);

Patchet, S.J. (1971); Cook R.F. (1974): Internal reports to Cleveland Potash Limited). The testing was carried out on cores taken from the exploratory boreholes. In the evaporite sequence, both the Upper Anhydrite and Carnallite Marl have significant differences in mechanical properties from the Potash and Halite rocks around them, with the Anhydrite being a relatively stiff rock and the Carnallite Marl having very little strength in either tension or compression. The difference between Potash and Halite is marked more by the inclusion of Shale in the Potash, where it can be up to 30% of the ore, and whereas Potash and Salt ores of similar Shale content have similar mechanical properties, the Potash ore is usually much weaker.

The Halite, Potash and Carnallite Marl all exhibit time dependant behaviour.

The Youngs Modulus and Poissons Ratio as presented in Table 2 are a considerable simplification of the behaviour of salt rocks because a proportion of the strain produced under load below the UCS is dependent on the stress used and how long it is applied. In the results in Table 2 the load was cycled three times and the third unloading cycle used to calculate the rock constants.

A plot of the UCS of Potash cores against width to height ratio, see Figure 2, shows that below a width/height of 1.5 the strength varies from 10 to 40 MPa. The UCS is related to Shale content, with an increase in strength proportional to decreasing Shale. Above a width to height ratio of 1.5 the UCS increases, demonstrating the effect of confinement on rock strengths.

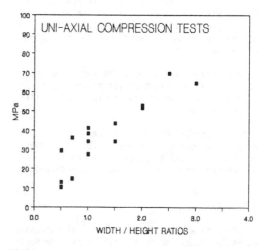

Figure 2

Table II

STRATA SEQUENCE	Thickness M	Depth M	S.G	E GPa	v	UCS MPa	UTS MPa
Lias/Drift	325	325	2.6	26.60	0.37	*	*
Rhaetic	18	343	2.5	13.20	0.05	*	*
Keuper Marl	251	594	2.7	34.60	0.08	*	*
Bunter Sandstone	293	887	2.5	27.72	0.10	*	*
Permian Marl	130	1017	2.3	27.17	0.07	85.77	6.86
Upper Halite	50	1067	2.2	22.06	0.24	30.10	1.59
Upper Anhydrite	8	1075	2.8	25.99	0.20	85.39	6.08
Carnallite Marl	15	1090	2.3	7.01	0.20	14.67	1.24
Potash	10	1100	2.1	22.42	0.37	30.13	1.79
Middle Halite	50	1150	2.2	23.66	0.28	26.77	1.63
Upper Magnesium Limestone			2.8	27.00	0.27		

* Not Tested

When the Uniaxial Compression Testing was carried out only a general description of each core was given and no detailed assay but the data used to compile Figure 2 does clearly show that for a similar strength, a high Shale ore does require a greater width to height ratio. Further testing would be interesting but experience underground bears this out.

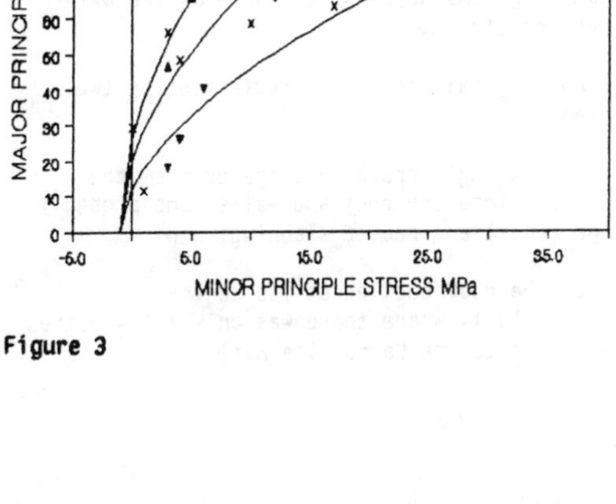

TRIAXIAL FAILURE ENVELOPE

Figure 3

4 TRIAXIAL FAILURE ENVELOPES

The triaxial testing, results of which are shown in Figure 3, shows that Halite develops more strength from confinement than Potash and Marl with the influence of Shale again reducing the effect of confinement on Potash, demonstrated by the wide variation in triaxial failure.

5 TIME DEPENDENT PROPERTIES

Using the Norton Power law $e = a\delta^n$ to represent the creep of the rocks, Table 3, where e is the strain rate and δ is the difference between major and minor principle stresses, gives the values of the constants a and n which provide a reasonable match to the triaxial creep tests.

The creep behaviour does not initially effect the yield pillar mining technique but its importance increases with time.

6 IN SITU STRESSES

Using the densities in Table 2, the vertical in situ rock pressure would be 22.377 MPa at 900 metres, 27.35 MPa at 1100 metres and 31.000 MPa at 1250 metres. A test to measure the in situ stresses using the overcoring technique gave a vertical stress of 30.1 MPa at 1100 metres and a

141

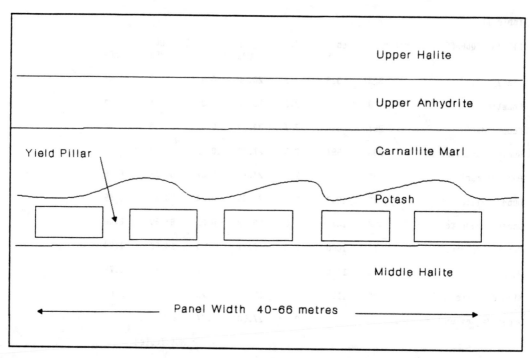

Figure 4   Cross Section of Yield Pillar Mining Method

Table III	a	n
Upper Halite	3 e-30	2.65
Carnallite Marl	3 e-30	2.80
Potash	3 e-30	2.75
Middle Halite	3 e-30	2.65

uniform horizontal stress of 15.4 MPa. (Hebblewhite, B.K. (1976): Internal report to Cleveland Potash Limited). These stresses were probably influenced by the tunnels around the test site, and, in an evaporite where creep does take place, the vertical and horizontal stresses should tend to equalise and I do not place any emphasis on these values.

7 FAILURE PATTERN AROUND A SINGLE OPENING

It was originally expected (1973 before mining commenced) that an "in seam" room and pillar layout with 6 metre wide rooms 3.5 metres high on 36 metre centres (extraction ratio 30.5%) would give good roadway conditions and create no damaging subsidence at either the base of the water bearing Bunter Sandstone or the surface.

Several variations in layout were tried, still using large pillars, but it became clear that Potash roadways mined on this system would not be stable in the long term and short term stability depended on local geological conditions. Potash seam thickness and Shale content were more variable than originally thought and it was not always possible to leave a consistently thick beam of Potash in the roadway roof without which the roof soon collapsed. Roadways which were required for long term travelling and conveying were then mined below the seam in the Middle Halite where they were more stable.

There was no damaging subsidence on the surface, however, and since there were no water inflows from the Bunter neither was there any damaging subsidence at the base of the water bearing strata.

Roadway failure was characterised by two events.

1. low angle cracks propagated from the corners into the roof and walls (and probably the floor) and caused slabbing; and

2. the roof buckled at its centre particularly where there was only a few metres of cover to the Carnallite Marl.

142

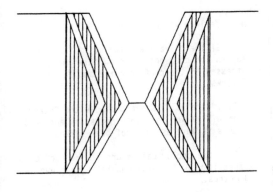

UniAxial Compressive Failure in Yield Pillar
Failed wedges are supported by Rockbolts

Figure 5

In the Halite, the first mode of failure was visible but the second, which developed in the Potash after six months or so was rarely seen.

Three reasons were put forward for this

1. The Halite contains less Shale than Potash and is generally stronger.

2. The effect of confinement is greater for Halite than Potash and hence strength increases more quickly with distance from the roadway.

3. Potash creeps more than Halite. The horizontal closure of a roadway is greater and it is this which causes the roof buckling, particularly where the roof beam is thin.

8 HYPOTHESIS OF MULTIPLE ROADS AND YIELD PILLARS

Experience gained in the Canadian Potash mines had shown that narrow yield pillars between rooms gave good roadway conditions and experiments were tried at Boulby. Initial trials were encouraging and a system of mining parallel rooms 6 - 9 metres wide separated by yield pillars 2 - 6 metres wide has evolved. A typical extraction ratio would be 72%, in a range of 62% to 77%, in the panel but with 66 metre wide panels separated by 100 metre wide rib pillars the overall extraction ratio remained at around 30%.

Initially, the yield pillar width was chosen on a width to height ratio of 1.5 to 2.0 on the assumption that what was required was to use the creep properties of Potash. It was found, however, that in order to achieve good roof conditions, the pillars actually had to

fracture and a situation created where a roof beam was supported on the failed pillars creating initial closure rates of 20 - 50 mm/day (See Figure 6). Extensometers in the roof showed that 95% of this movement occurred also at the top of the seam regardless of seam thickness. Several attempts have been made to drill through the Carnallite Marl and install instrumentation in the Upper Anhydrite but it is a weak rock and has a tendency to squeeze and grip the drill rods. None of these attempts were successful and it is not known exactly how the Carnallite Marl, the Upper Anhydrite and Upper Halite behave. However assuming the Potash lowers and separates from this Carnallite Marl then the Marl would have to span the opening, 40 - 66 metres wide, or collapse.

Observations of the Marl where it has been exposed in the roadway roofs shows clearly that it cannot span even 6 metres and it will certainly collapse.

A plane strain bending moment analysis on the Anhydrite shows that it requires a tensile strength of 2.25 MPa to span 66 metres and it seems likely that in large areas of the mine it may be able to span the panel though it will bend.

In terms of a 66 mm span the Upper Halite can be considered "massive" and is in fact the panel roof.

The Potash beam therefore is supported on the yield pillars and loaded by the Marl, possibly the Anhydrite, and horizontal stress caused by the creep of the rib pillars towards the panels.

It is important to keep the yield pillar the correct strength. Too weak and the roof is not supported, giving high closure, too strong and the roof fails and requires repair and support which is costly and delays production.

The essential feature of the mining technique is to isolate the roadway roof as much as possible from the high stresses at this depth. In order to achieve this the roadway roof (Potash, Marl and Anhydrite) must be lowered away from the panel roof (Upper Halite) and this is achieved by allowing the yield pillars to fail in Uniaxial Compression. The only force then acting on the roadway roof is the high induced horizontal stress. By mining at least four roadways, the flank roadway roof can be allowed to fail isolating the centre roads from all high stresses. It is an important feature of the method that the roads should be advanced as close together in time as possible with the flanks leading the centre roads by only one

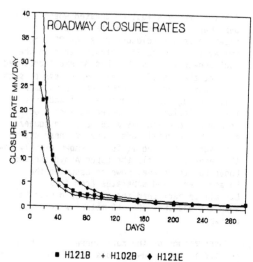

H121B + **H102B** ♦ **H121E**

Figure 6

mining cycle. This ensures that the yield pillars are created and can then fail.

9 OPERATION OF YIELD PILLAR TECHNIQUE

Pillar Strength

Actual Pillar width is dependant on

1. Potash strength
2. roadway height
3. difference in elevation of adjacent roads
4. use of blasting or continuous mining

In strong ore, a pillar with a w:h ratio of 0.75 is usual and in the weakest ore this rises to 1.5. Blasting tends to damage the pillars and in a strong ore which is mined with continuous mining the w:h ratio would come down to 0.65. Where there is a dip across the panel and adjacent roads are mined at different elevations the w:h ratio may be reduced to allow for the different shape of the pillar.

Assuming 2 metres of Potash and 15 metres of Marl, the "Dead Load" on the failed pillars would be 0.5 MPa. Including the Anhydrite, this would increase to 1.3 Mpa but this is an upper limit since some load would be held in shear at the abutments. The yield pillars thus support the ground and any buckling load, by continually yielding if necessary, and to maintain good roadway roof conditions the Potash beam only has to carry these loads over the 6 - 9 metre roadway span.

There are basically three types of roadway

beam failure which now occur

1. If the pillars are too stiff, and bed separation does not occur quickly enough stresses in the roof increase and failure follows.

2. If the pillars are too weak the Potash beam will lower more than necessary and this causes failure as the roof beam flexes.

3. If the geological nature of the roof beam makes it weak it maybe unable to span even 6 - 9 metres.

Because the forces are low, these problems can usually be dealt with by chocking for support and only roof failure which falls into category 3 is likely to prevent mining. Roof failure, however, is costly both in the use of labour and materials to support the roof and in the loss of production time. In order to try and keep this to a minimum one man is employed continuously monitoring the closure rate of the panel roadways and examining the yield pillars. Too high, or too low, a closure rate is investigated and if the pillars are the wrong size changes are made. In addition, a continuous check on the ore type is made and if it changes the pillar width is changed as necessary.

One solution to the problem of pillar strength would be to replace the pillars with some kind of pack of known strength and have the advantage of 100% ore recovery in the panel at the extra cost of the packs. The failure due to weak roof beam and the associated repair costs would still be there but there would be less failure due to weak or strong pillars. However, due to the low value of the ore and the relatively high costs of packing materials this is not economic.

The usual Potash roadway life is one year but this is a function of ventilation distance and air leakage and not roadway stability. During this period, the roadway closure rate continues to decline provided no mining occurs within 120 metres or so of the panel. If mining does approach within this distance, the closure rate increases until mining ceases and then declines to slightly more than the original levels (See Figure 7). This increase does not cause any operational problem but it does indicate that higher overall extraction ratio would have a measurable effect on roadway roof stability and we are currently using mathematical modelling to evaluate whether we can increase the extraction ratio without causing worse mining conditions or damaging strains at the base of the water bearing strata above.

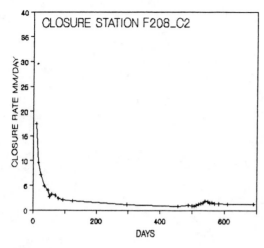

Figure 7

Mining Engineer Vol 141 March, 517-522.
CHILTON, F & ROBINSON, G (1984) Forcing
Ventilation System at Boulby Mine, Cleveland
Potash Limited, (Technical Note). Trans.
Insts. Min. Metall. (Section A, Mining
Industry) Vol. 93, A82-83.

REFERENCES

WOODS, P J E (1973) Potash Exploration in
  Yorkshire, Boulby Mine Pilot Borehole. Trans.
  Instn. Min. Metall. (Section B; Appl. Earth
  Science) 82 B99-106.
CLEASBY, J V et al (1975) Environmental aspects
  of Boulby Mine. Cleveland Potash Ltd.,
  Yorkshire. Minerals and the Environment
  (London: IMM) 691-713.
CLEASBY, J V et al (1975) Shaft Sinking at
  Boulby Mine, Cleveland Potash Ltd., Trans.
  Inst. Min. Metall. (Section A; Mining
  Industry), 84, A7-28.
PEARSE, G E (1975) The Selection of Mining
  Equipment and the Initial Operations at
  Boulby Potash Mine, Cleveland Potash Ltd. The
  Mining Engineer Vol. 134, Aug/Sept 585-594.
WOODS, P J E (1979) The Geology of Boulby Mine,
  Cleveland Potash Ltd., Econ. Geol,. 74, 409-
  418.
WOODS, P J E and HOPLEY, R J (1980) Horizontal
  Longhole Drilling Underground at Boulby Mine,
  Cleveland Potash Ltd. The Mining Engineer Vol
  139, 22, 585-591.
McPHERSON, M J & ROBINSON, G (1980) Barometric
  Survey of Shafts at Boulby Mine, Cleveland
  Potash Ltd., Cleveland. Trans. Instn. Min.
  Metall. (Section A; Mining Industry), 89,
  A18-28.
CLEVELAND POTASH LIMITED The first UK Potash
  Mine begins production The British Sulphur
  Corp. Ltd.
ROBINSON, G et al (1981) Underground
  Environmental Planning at Boulby Mine,
  Cleveland Potash Ltd. Trans Instn. Min.
  Metall. (Section A, Mining Industry) 90,
  A107-121.
CHILTON, F & LAIRD, K L (1982) Problems
  encountered in working potash at depth. The

Effects of Geomechanics on Mine Design, Kidybiński & Dubiński (eds) © 1992 Balkema, Rotterdam. ISBN 90 5410 040 0

A method for forecasting of the gasodynamic events in Nowa Ruda coal mine

J. Dubiński
Central Mining Institute, Katowice, Poland

A. Dybciak & J. Jeżewski
Hard Coal Mine 'Nowa Ruda', Poland

ABSTRACT: A multi-parameter method for forecasting of gasodynamic events which takes into account gasodynamic, geomechanical and geodynamic factors, has been developed. A value of the complex index of outburst hazard **KWZ** allows establishing **4** stages of hazard from low to high. Results of testing the method under the conditions of coal mine **Nowa Ruda** show a visible increase in effectiveness of the forecasting as compared with the existing methods based only on gas parameters.

1 INTRODUCTION

The Nowa Ruda coal mine conducts mining operations in a hard coal deposit situated in the south-eastern part of the Lower Silesian Coalfield (see Fig. 1). The occurrence of gas and coal outbursts hazard during the mining of seams that are highly saturated with deposit gases is a specific quality of this coalfield (Cis, 1971).

A particular intensity of this phenomenon occurs in the part of the mine called "Piast panel". In this panel the first outburst took place in 1908 and since that time 1278 gas and coal outbursts and 10 sandstone and CO_2 outbursts have occurred. In one of the greatest outbursts, with more than 5000 tonnes of displaced material, more than 700 000m^3 of CO_2 were forced into the net of workings. Also mine disasters have been associated with outbursts the greatest of which occurred in 1941 and took a toll of 187 human lives.

Hard coal deposit of the Piast panel consists of Carboniferous strata overlying the crystalline basement. Coal measures as a series of balance coal seams with thicknesses of from 1m to as far as 7m lie both within the shale complex and amidst the overlying thick sandstone and conglomerate layers. This series is immediately underlain by the so called Nowa Ruda facies composed of argillites and ferruginous or refractory arenaceous shales with coal intercalations of varying thickness (see Fig. 1).

Distinctive features of this deposit are as follows (Najsznerski et al, 1988):
- high carbon dioxide saturation confirmed by gas-bearing capacity of more than 20m^3 of CO_2 per tonne,
- low compaction of coal of from 0.18 to 0.60 on the Protodiakonov scale,
- complicated tectonic structure,
- great depth of mining (700-800m),
- monoclinic dipping of strata at an angle of about 25°.

2 CHARACTERISTIC OF EXISTING METHODS USED FOR THE ASSESSMENT OF OUTBURST HAZARD

As mentioned above, first indications of outburst hazard in the Piast panel appeared in 1908. At first, these were only small events occurring sporadically (Szewczyk et al, 1987).

As a depth of mining was increasing, the observed outburst hazard grew higher both in number and intensity as defined by a quantity of material thrown into

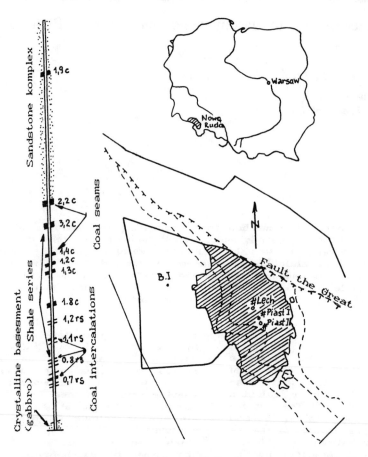

Fig.1. Positional and structural characteristic of the coal deposit in coal mine **Nowa Ruda**

workings. It is shown on Fig. 2 and Fig. 3. To control the hazard, a joint action must be taken which includes combined elements of prediction and prevention (Kidybiński 1982).

This paper is aimed at presenting a complex method for forecasting of gasodynamic events, which is one of the latest achievements in this field.

At an early stage of development of methods for assessing the outburst hazard only superficial observations of the events associated with the increase in hazard had been used, such as cracks, sidewall spalling, increased quantity of gas produced from a rock body, effects during the drilling of boreholes, etc.

At the next stage, these observa-

tions were transferred to the test boreholes drilled for the assessment of outburst hazard. First measurements of this type were introduced in coal mine Nowa Ruda in 1927. They were based on measuring gas pressure and temperature in test boreholes of 2 to 3m deep. The first criterion for outburst hazard was the gas pressure in the test boreholes exceeding 1 at. At that time the influence of disturbances in regularity of coal seams as a hazard factor was also taken into consideration. In those days, many hypotheses for the mechanism of the outburst phenomenon had come into being, which were based on gas theories as for instance, nest theories, reservoir theories, etc. However, further increase in depth of mining (see Fig. 3) has qualitatively changed the level of hazard. Practically, at depths below 400m, the effectiveness of the above outburst hazard assessment began to decline. Later works on development of the assessment methods were based on increasing the density of gas measuring points, modification of the hazard level and on implementing the measurements of desorption of gases from borehole coal cuttings. At first, this improved the assessment effectiveness, but as the depth of mining was increasing, especially in the years sixties and seventies, there also increased both the outburst hazard and the drop in the effectiveness of the method. At that time the growing share of the stress factor in the hazard development was noticed more and more. This gave rise, among other things, to taking into account the interaction of all the exploitation edges as well as the stress distribution in sidewall zones. From it there emerged several hypotheses for the outburst mechanism which ascribed the basic role to stresses and physico-mechanical properties of the rock mass. In extreme cases the outbursts were even identified as rockbursts.

In 1973 in coal mine Nowa Ruda an attempt was made to enclose both main gas parameters and stress state in the evaluation of the outburst hazard. During the drilling of the test borehole with a specially made device (the so called energy probe) an energy balance of gas, stress state and the relation between these factors were determined by measuring the following indices (Krzemiński & Górkiewicz, 1974; 1975) :

-gas energy index E_G which characterizes the energy of gas released from a given amount of coal following the coal body disintegration,
-drilling work L_W defined as the coal effective strength index,
-index V_P for the amount of gas released during the drilling of coal with an open chamber of the probe.

No doubt, this method has given an impulse to searching for new trends of prediction that would allow for changes in the stress state of the coal body. However, the method lacked quality that would allow using it in mining practice. Therefore, the ideas enclosed in it began to find applications only by the end of seventies that is, at the time when geophysical methods were introduced to underground mining on a larger scale. Also, in the Nowa Ruda mine these methods as auxiliary ones have found application to the immediate forecasting of the outburst hazard.

3 FUNDAMENTALS OF THE COMPLEX METHOD FOR FORECASTING OF GASODYNAMIC EVENTS

From the scientific and practical experiments on the assessment of outburst hazard, it has been found that the acceptance of a multi-parameter outburst model considered as the energy phenomenon in the course of which complicated processes of phase changes in the rock structure and the saturating gas take place is justified. This complex multi-parameter model defines a new approach to the problem of outburst hazard prediction. Because of many measuring points and their variability in time and space the problem of complex and fast processing of measurements data files could only be solved by the computer technique. This condition in the case of the Nowa Ruda coal mine, was fulfilled by the end of eighties.

The afore-mentioned technical possibilities of the fast processing of the large data files, the acceptance of the hipothesis for the multi-parameter and energy model of the outburst phenomenon, technical possibilities of conducting various types of measurements in mine workings [as for example, seismic measurements of propagation velocity of a seismic wave ahead of a longwall face (Dubiński et al, 1985) - see Fig.4] which extend the range of deep penetration of the coal body, have led to the origination of a production-immune complex method for the assessment of gas and rock outburst hazard. Its essence consists in grouping of measurement and observational data into three following groups of factors :

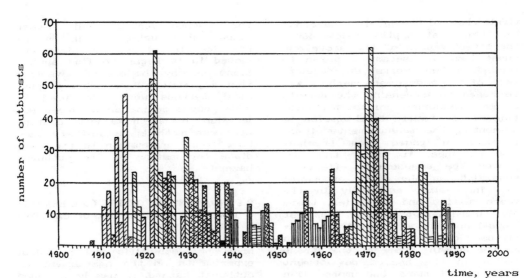

Fig.2. An average quantity of outburst material and a number of outbursts in coal mine **Nowa Ruda** during the period of 1900 – 1990.

▨ <100t, ▧ 100-200t, ▨ 200-500t, ▥ 500-1000t, ▤ 1000-2000t, ▦ 2000-3000t, ■ >3000t

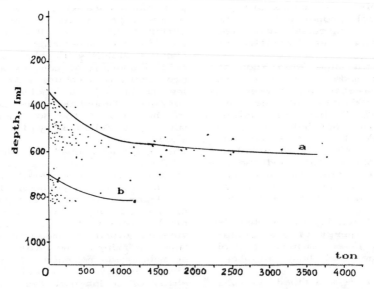

Fig.3. Dependence of the outburst quantity on depth under the conditions of coal mine **Nowa Ruda**
a – in the case of mining coal seam No 410/2+412
b – in the case of mining the coal intercalation in a layer of refractory shale

-**Gasodynamic factor GZD** comprising information about deposit gas parameters such as the type of gas (g), overpressure (p), outflow of gas from test boreholes of from 3 to 6m deep (Q), and desorption of

150

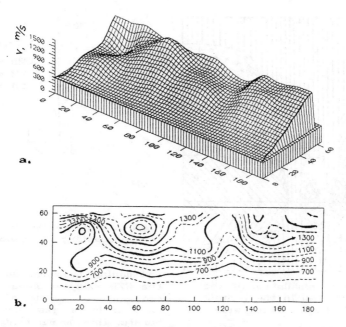

Fig.4. Velocity distribution of seismic longitudinal wave
in a coal seam ahead of longwall No 103 (1991-01-26)
a - velocity stereogram, b - velocity map

gas from coal cuttings collected in the course of drilling these boreholes (Δp). Depth distribution of gasodynamic events occurring during the drilling of test boreholes of 15m deep, which characterizes intensive, hypernominal flow - off of cuttings (W); traces of gas (Slg) or emanation of a larger quantity of gas along with cuttings; and the blow-out of gas and cuttings (Wyd), which has the quality of a minioutburst from a borehole are also taken into account.

-**Geomechanical factor GM** describing such parameters as the thickness of a layer under outburst hazard (m), compaction of outburst prone rocks (fz) measured by the Protodiakonov method, regularity of strata deposition (geo), dip of strata (α) and dip of the working (β).

-**Geodynamic factor GD** is the set of parameters describing the local stress distribution resulted from both depth of mining (gl), effects of exploitation edges (k) and a degree of rock mass destressed by an earlier mining (odp). The effect of these elements is de-

termined by means of the set of simulator programs based on both the results of research works and seismic measurements conducted by the mine. These programs take into account the effects of such factors as the thickness of a mined out panel, spatial distribution of the effect of an edge of stopped mining, the support zone of caved area and the rheological factor. An example of such a stress distribution in the vicinity of a longwall face is shown on Fig.5.

Each factor, **GZD, GM, GD**, is individually processed in accordance with the general law which states that they are multi-parameter systems where every parameter will be scaled according to a separate calibration curve. Each curve has been developed and tested under the real conditions of a wide range of cases of the formation of the gas and rock outburst hazard in a mine. Therefore, the aforementioned factors can now be presented in the following forms :

GZD = f(g,p,Δp,Q)+f(Wz,Slg,wyd)

GM = f(m,fz,geo,α,β)

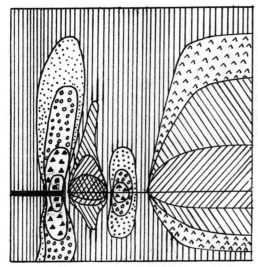

Fig.5. Computer simulation of the stress distribution in the vicinity of a longwall face worked with caving.

▨ gob, ▨ structural strain, ▧ partial strain, ▲▲ subsidence, ▥ normal stress, ▦ raised stress, ▨ medium stress, ▲▲ high stress, ▥ maximum stress anomaly

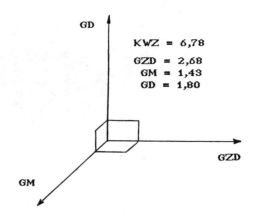

KWZ = 6,78
GZD = 2,68
GM = 1,43
GD = 1,80

Fig.6. A multi-parameter system for forecasting of gasodynamic events

$$GD = f(gl,k,odp)$$

This means that the value of each factor is computed as a multi-parameter system independent of the remaining factors. According to such an approach each one, at a given moment, determines its share in a general assessment of the state of outburst hazard whereas complex hazard index **KZW** gives a total level of the hazard created by gasodynamic events. Under this notion the following dynamic events are under-stood :
- gas and rock outbursts,
- outflow of liberated gas,
- mining tremors,
- rockbursts,
- intensive rock slides,
- rapid closure of workings, etc.

Therefore, when considering a spatial system where coordinate axes correspond to respective factors **GZD, GM, GD** (see Fig.6), it can be possible to assess the occurrence of any of these events.

For example, on a plane with axes **GM** and **GD** there are determined such dynamic events as rockbursts and mining tremors. On introducing the **GZD** axis we obtain the extended range of events which additionally includes gas and rock outbursts. This means that the aforementioned events have a common origin, i.e., they results from multi-phase energy changes that take place in the rock medium.

It should be noted that gas and rock outbursts are more widely understood dynamic events and that such event as the rockburst is a form of outburst without the participation of the gas factor. This means that the rock skeleton in which, because of its collecting properties (porosity, permeability etc.), a gas could be accumulated in a free, sorbed or liquid phase, would become endangered by gas and rock outbursts under the specific stress and strength conditions. These rocks as a rule, have low compaction and high porosity. It has been assumed that the gas - especially in its liquid phase - filling out the pores and fissures of the medium makes the structure stiff causing it to be a typical equivalent of the most compact bursting strata. In the course of outburst the destruction of the rock structure progresses gradually as opposed to the rapid process of rock destruction during the rockburst. It develops as the destructed material is being transported by a highly desorbing gas until the stress equilibrium is reached. Thus, the outburst is an event extended in time (from seve-

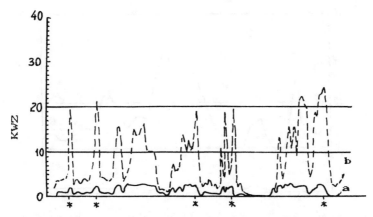

Fig.7. The assessment of the hazard state in tailgate of longwall No 301 in bunch 4 of refractory shale
a – from the existing indices of hazard (p, Δp_2),
b – from the complex indices of hazard **KWZ**

ral seconds to even a dozen or so of minutes). Of course, the degree of outburst hazard depends on minimum boundary values of individual gasodynamic parameters. This is confirmed by the fact that different gases and rocks participate in the outbursts and that the minimum depth of the first outbursts that occurred in various mines is different. But the mechanism of these phenomena is the same and from the presented methodical approach to the forecasting of outburst hazard results the versatility of the worked out complex method. The values for parameter **KZW** have, here, energy dimensions expressed in conventional units.
From the test material obtained in coal mine **Nowa Ruda** under diverse conditions the scale for assessment of the hazard has been worked out :

0 < **KZW** < 10 Low state
10 < **KZW** < 15 Raised state
15 < **KZW** < 20 Small gasodynamic
        events possible (initial
        phase of hazard)
20 < **KZW**   High state of out-
        burst hazard.

4 RESULTS OF THE HAZARD FORECASTING IN COAL MINE NOWA RUDA

The existing experiments of testing the presented method for fore-

casting of outburst hazard prove its high effectiveness. This statement can be illustrated by the results of several chosen examples obtained in coal mine Nowa Ruda.
The first one refers to the forecasting for the tailgate of longwall No 301 being driven at a level of -320m in bunch 4 of the refractory shales. The distribution of parameter **KZW** and the outburst hazard indices ($p=0.3$ at; $\Delta p=120mm$ H_2O) used hitherto, by mines (see Fig 7) shows clearly a much higher resolution and sensitivity of the **KZW** parameter curve. On the horizontal axis of the graph there are marked places where the outbursts or the other gasodynamic events have occurred.
Similar findings have been provided by the analysis of the **KZW** parameter distribution (see Fig.8) obtained during the drivage of the tailgate of longwall No 103 in a partly destressed seam of No 415/2 at the level of - 260m. Distributions of respective outburst hazard parameters **KZW**, p, Δp_2 obtained in the course of mining longwall No 103 are shown instead on Fig.9. A higher effectiveness of the complex method for forecasting by using parameter **KZW**, also in the case of conducting the exploitation by means of the longwall system, is here clearly visible. Particularly visible on the **KZW** parameter graph are the elements of

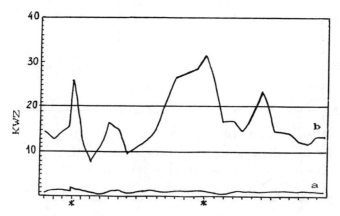

Fig.8. The assessment of the hazard state in tailgate of longwall No 103 in coal seam No 415/2
a – from the existing indices of hazard $(p, \Delta p_2)$,
b – from the complex indices of hazard **KWZ**

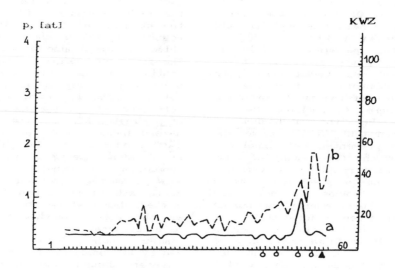

Fig.9. The assessment of the hazard state in longwall No 103 in coal seam No 415/2
a – from the existing indices of hazard $(p, \Delta p_2)$,
b – from the complex indices of hazard **KWZ**,
▲ – the zone where the working face is passing the edge of stopped mining in adjacent seams.

passing with mining operations the following areas :
- exploitation edges in other seams,
- post-outburst caverns,
- remnants and coal pillars,
- zones of geological disturbances.

The results of correlation between the seismic activity of a rock mass and the averaged values of parameter **KZW** determined for long-wall workings indicate that this activity is closely connected with the development of outburst hazard.

5 CONCLUSIONS

1. The complex method for forecasting of outburst hazard has been based on the assumption that the gas and rock outburst is a physical phenomenon resulted from the complex energy processes that are taking place within the rock skeleton saturated with a gas being in various phases of the state of concentration (sorbed, free and liquid). The gas factor is here an additional element of the accumulation of energy.

2. The complex method for forecasting of outburst hazard consists in a multi-parameter analysis that takes into account gasodynamic, geomechanical and geodynamic factors. The role of those last-mentioned factors is clearly rising with the increase in depth of mining.

3. The examples of mining measurements show that the worked out method for forecasting of outburst hazard is evidently raising the prognosis effectiveness as compared with the existing methods which are only based on gas parameters.

4. Obtained experiences indicate that the future development of the method for forecasting of outburst hazard should be aimed at the extension of the rock mass deep penetration range and at allowing the evaluation of the energy accumulated in the vicinity of a studied working, including, at the same time, the determination of its near face zone response to an excessive energy.

REFERENCES

Cis, J. 1971. Gas and rock outbursts in the Lower Silesian Coal field, Katowice, ed. Śląsk

Dubiński, J. Szot, M. Dybciak, A. Komandowski, P. 1985. Application of geophysical methods in the problems of gas and rock outbursts, Publ. Inst. Geophys. Pol. Acad. Sc.M-6(176), ed. PWN.,Warszawa-Łódź.

Kidybiński, A. 1982. Principles of mining geotechnics. Katowice, ed. Śląsk.

Krzemiński, T. Górkiewicz, P. 1974 The energy state of gas and coal in the light of a new method for measuring indices of the gas and rock outbursts hazard, Przegląd Górniczy. No-5.

Krzemiński, T. Górkiewicz, P. 1975 A new method for determination of the coal drilling work in seams as an index of stresses existing in the sidewall of workings. Arch. Górnictwa PAN, No 3.

Najsznerski, R. Stalski, L. Wróbel L. Dybciak, A. 1988. Formation of the outburst hazard and evaluation of the preventive measures in "Piast" coalfield of Nowa Ruda mine. [in]: Tendencies in gas and rock outburst hazard prevention in underground mines. Radków-Nowa Ruda.

Szewczyk, K. Kaleciński, K. 1987. Evidence of gas and rock outbursts in the Lower Silesian hard coal mines. Proc. of the Commission on gas and rock outbursts, No 9, Wałbrzych.

Effects of Geomechanics on Mine Design, Kidybiński & Dubiński (eds) © 1992 Balkema, Rotterdam. ISBN 90 5410 040 0

New concept of destressing in outburst prone rocks

Z. Rakowski
Mining Institute Ostrava, Czechoslovak Academy of Sciences, Czechoslovakia

ABSTRACT: The problem of very intensive outbursts of sandstone and CO_2 in Slany shafts showed on the necessity to find new concept of destressing. The new concept is based on large scale specific fracturing of the rock below the shaft floor. Results of mathematical modelling theoretically confirmed expected principles of the idea. The hydrofracturing and water jet technology seem to be suitable for practical application of the concept.

1 INTRODUCTION

The problem of very intensive outbursts of rocks and gases belongs among the problems which have not been solved yet satisfactionally and that occure during various types of mining activities. A number of such outbursts of sandstone and carbon dioxide have occured during sinking of two deep shafts in Slany coalfield near Prague, the capital of Czechoslovakia. The character of these phenomena, namely their intensity and the scale of works which have been realized to prevent their occurrence provide interesting materials for better understanding of the mechanism and the ways of prevention of the outbursts.

2 THE SITUATION

Two shafts (no.1 and no.2) were sunk in the Slany coalfield as the first mining works entering new part of Kladno coalfield which is a part of Central Bohemia Coal Basin of carboniferous age. Due to the great depth of coal seams the total depth of the shafts was designed up to 1200 m below the surface. The diameter of both shafts was the same and was 8.5 m (internal) and 10.1 m (total) respectively. The distance between two shafts was about 60 m. Both shafts were sunk classically by blasting and lined by simple monolite concrete. Finally the shaft no. 1 achieved the depth of 1006.7 m and the shaft no.2 the depth of 912.2 m so that no one was sunk up to planned depth. Extremely difficult technical and technological problems with outbursts resulting in very high costs of works were the main reason for stopping the shaft sinking. The whole original concept of mining activities in the coalfield was interrupted, too (Fig. 1).

From geological point of view the so called "Mirošovice beds" was the most important body. "Mirošovice beds" are represented by about 250 m thick interval of sandstones and conglomerates beginning at the depth of about 800 m below the surface. It means that shaft no.1 passed through the whole thickness of this part and entered coal bearing strata. On the contrary the shaft no.2 had to be stopped even before passing through these beds with actual depth of 919.2 m.

"Mirošovice beds" are not some homogeneous or monolitic body. It is rather a set of beds of sandstones and conglomerates of medium

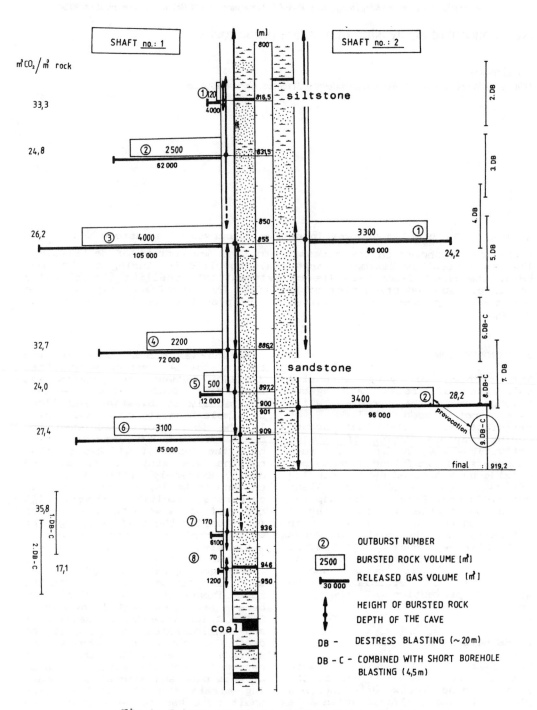

Fig.1 Scheme of Slany shafts with outbursts

thickness divided by thin layers of siltstones (Fig. 1). The sandstones have specific structure characterized by very poor cementation with many pores and direct contacts of mineral grains.

These structural features influence physical and mechanical properties of such a type of rocks very remarkably. Thus the sandstone has rather high porosity varying between 10% and even 20%. Its uniaxial compressive strength is quite low and varies in the range of 10-20 MPa. Tensile strength was practically impossible to measure due to high friability and brittleness of the material. It could be only estimated that it would be of minimal value very near to zero. Exploration works prior to the shaft sinking revealed that porous system of "Mirošovice beds" is filled by pressurized and highly mineralized water. Therefore, quite large scale injection of cement material trough surface boreholes were done to "Mirošovice beds" prior to the shaft sinking. It is very likely that this operation contributed also to lower permability of subjected rocks varying between 10^{-7} and 10^{-9} m^{-1} according to in situ measurements. Thus even rather dense and irregular net of macrocracks did not influence the permeability remarkably. High stresses resulting from the depth also contributed to permeability conditions. Sporadic overcoring measurements of stress field showed on increased horizontal stresses even equal or two times greater than vertical stresses. However, we must consider these results as rather approximate due to specific mechanical properties of the rocks - mainly low strength and high brittleness - which surely influenced the accuracy of these measurements. On the other hand an indication of somehow but reasonably increased horizontal stresses was done and has to be taken under consideration during further analysis.

3 THE MAIN FEATURES OF THE OUTBURSTS

Ten outbursts of sandstone and carbon dioxide occured during the sinking of Slany shafts. In the shaft no. 1 which was sunk with about 120 m advance ahead the face of the shaft no. 2 eight outbursts took place. In the shaft no.2 only two cases were registered. General scheme of outbursts occurrence in both shafts is presented on Fig.1. The first outburst was unexpected and occured when the floor of the shaft no.1 was at the depth of 816.5. The intensity of the outbursts was variable. The amount of bursted rock material varies from 120 m^3 to 4000 m^3 per one phenomenon. Adequately great differences in the volume of gas released by individual phenomenon were observed. The most intensive outburst released about 105 000 m^3 of CO_2 to the shaft. On the other hand only 2000 m^3 of gas was bursted to the atmosphere by the smallest outburst. So the specific amount of the released gas to the bursted volume of the rock means the range of 17.1 to 33.3 m^3 of CO_2 per one m^3 of bursted sandstone. Expelling the value 17.1 belonging to the outburst no.8 in the shaft no.1 - which was visibly influenced by destress blasting - the specific gas content is about 28.5 m^3 CO_2/m^3 of bursted sandstone. Taking under consideration some bulk coeficient (approx. 1.9) it gives more than 50 m^3 of CO_2 per m^3 of solid rock. But the vicinity of after burst caving is partly degasified by the outburst, too. The affected rock volume could be of the same order as burst cavity. By such a consideration we can estimate the specific gas content of intact rock as about 25-40 m^3 of CO_2 per the same rock volume unit. It could means that gas pressure in the intact rock is within the range of 5-8 MPa. This number although rather roughly estimated from field observations has some importance namely due to practically unsuccesful direct gas pressure measurements.
There are some further important features of Slany burst phenomena:

- 9 outbursts were iniciated by normal blastings which represent about 400-750 kg of explosives per one blasting operation
- 1 outburst was iniciated by destress blasting

- the whole sequence of "Mirošovice beds" in the shaft no. 1 was affected by bursts, in the shaft no.2 the first outburst occured at the depth 855 m
- the greater thickness of sandstone the more intensive outburst occured
- bursted material is formed practically by individual mineral grains or even their broken parts so that original structure of the rock was totally degradated
- the cavings after the bursts were more or less oriented vertically downward and were filled by bursted rock material
- the sandstone in the vicinity of the caving was always characteristically fractured, forming the so called "onion-like structure"
- consisting of dense set of fine rock layers
- almost the whole thickness of appropriate sandstone bed was in fact affected by the outburst either by caving or by specific fine "onion - like" type of fracturing.

4 THE MECHANISM OF THE OUTBURSTS

It is necessary to formulate some imagination about the mechanism of phenomena to find some effective measures against them. For specific case of Slany shafts we can describe the bursting mechanism as follows:

- Great volume of the pores, high gas pressure and high rock stress cause that "Mirošovice" sandstones should be considered as the formation with high specific internal energy.
- High brittleness and low tensile strenght of sandstone make it prone for rapid changing of mechanical state as the result of sudden changing of boundary conditions.
- Removing of the certain volume of the rock by blasting for new advance of the shaft face leads to sudden forming of new open surface of the rock - new boundary.
- There was a state of equilibrium on this boundary before blasting. The non-equilibrium state instantaneously occures on the boundary

formed after blasting.
- Due to high internal energy the conditions for forming the dynamic stress drop on the boundary are fulfilled. The combination with low tensile strength and high brittleness anables to generate the bursting of thin layer of the rock from newly formed surface. By repeating this situation the so called burst wave develops and spreads to the rock mass.
- "Onion like" type of fracturing is the result of burst wave performance.
- Thin rock pieces are explosively desintegrated to mineral grains by the expansion of gas captured in the pores.
- Fine mineral grains or their parts are pneumaticcally transported to underground space.

With some simplification outburst mechanism could be described by such a definition:

The outburst of rock and gas is a series of instantaneous static events driven by tensile failures generated by rapid drops of internal energy on the plane which spreads to the rock mass like rarefraction shock (burst) wave forming the cavity.

5 THE DESTRESSING BY LARGE SCALE BLASTING

From the above described mechanism of the outbursts it is quite clear that it is necessary to decrease internal energy to avoid iniciating these phenomena. In Slany case large scale of destress blasting was used for this purpose. Some typical scheme of such a blasting is showed on Fig. 2. The main idea of this type of destressing is to form a system of fractures by blasting a group of long boreholes and by this to anable the gas release from the rock through free boreholes to the underground atmosphere. The typical parameter are as follows:

the length of boreholes ... 17-21 m
charged interval .... 8-10 m
explosives per borehole ... 40-65kg
explosives per blasting. .445-950kg

In some cases short boreholes were

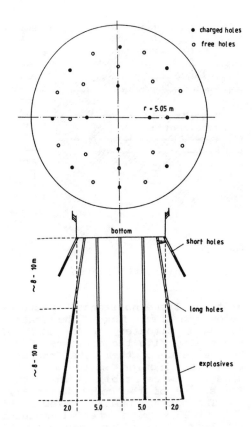

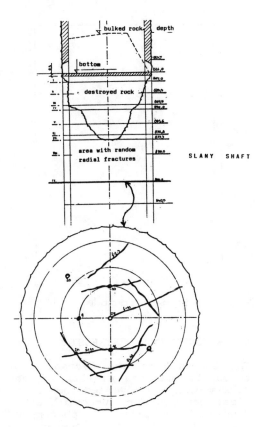

SLANY SHAFT

Fig.2 Scheme of destress blasting   Fig.3 Effects of destress blasting

blasted together with long ones, too. In shaft no. 1 only two destress blastings were realized (see Fig. 1). The other part of dangerous rock interval was so destroyed by outbursts that it was impossible to use any way of destressing at all. In shaft no. 2 nine blasting operations were realized altogether. The last one provoked the outburst at the depth of 900 m (Fig. 1).

Fig. 3 shows the scheme of effectivness of destress blasting. There were three main zones of different effect of destress blasting:

- bulked rock material on the floor (1-3 m high),
- rather small "core" of conic shape of destroyed rock below the shaft floor,
- larger area with random radial and not persistant fractures in deeper part of blasted interval.

Gas releasing ratio was rather low and varied from zero to 340 m^3 of CO_2 released per one blasting.

The occurence of intensive outbursts was not significantly affected by used destress blasting. The explanation is based on the following conclusions:

- in fact porous gas collector system and gas pressure were not affected,
- the system of random newly formed fractures had only small effect on rock stress field,
- radial vertical fractures remained closed by high horizontal stresses and so they were not permable for gas.

Thus, principal factors influencing the iniciation and the mechanism of the outbursts were not significantly affected by such a way of destressing.

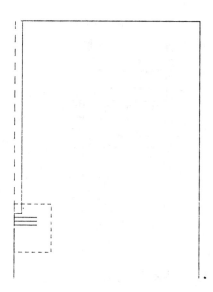

Fig.4 Axisymmetric model

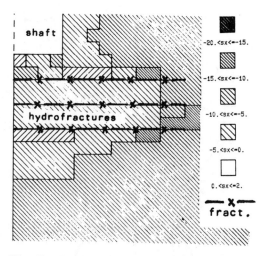

Fig.5 Horizontal stresses

6 NEW CONCEPT OF DESTRESSING

The analysis of the mechanism of outbursts and the effects of de-stress blasting revealed some prin-cipal demands for succesful avoi-ding such an intensive phenomena. First of all it is necessary to minimize stress drops occuring on the new boundaries formed by advan-ce of the shaft. Significant volu-metric destressing in comparatively large area arround the bottom of the shaft should decrease internal energy in the rock mass. More smooth way of entering the rock mass by the shaft is demanded for minimizing shock effects on newly formed boundaries, too.

According to the new concept demanded destress effect is achi-eved by a system of artificial fractures which are approximately parallel to the shaft floor and form several rows below it. The fractures should have the orienta-tion and scale roughly similar to these paterns of fracturing known from described burst mechanism. The main idea is in the atempt to adopt the mechanism of destressing to the mechanism of natural phenomena as much as it is possible in each specific case.

Stress analysis by FEM was done by using axisymmetrical numerical model and simulating the influence of three large horizontal fractures below the shaft floor. The fractu-res were situated in the axis of the shaft, with the distance of 2 m between each of them. The fractures were of circular shape with the diameter twice as big as shaft one. The scheme of the model is on Fig. 4 The package GEM22 - Rakowski et al. (1988)- with special modulus for contact problems -developed by Dostál (1990)- for modelling of the fractures was used. The stresses were calculated using isotropic linear elastic solution.

The results of the modelling could be presented e.g. by changes in horizontal stresses and maximum shear stresses (see Fig. 5 and 6). Remarkable decreasing of both types of stresses are visible in the part of the massive influenced by hori-zontal fractures. The most probable explanation is that the fractures are oriented in preferable directi-on of shear stresses and so quite a big displacement could occure on them. The concentration of higher stresses is visible behind the zone of fracturing, of course. But it is rather far from the shaft lining. The differencial movement along the fractures could lead to partial opening of them in vertical direc-tion. Remarkable decreasing of stresses in affected area could be very helpful for reopening of natu-ral system of fractures which exist there and anabling quite reasonable gas releasing from the rock mass.

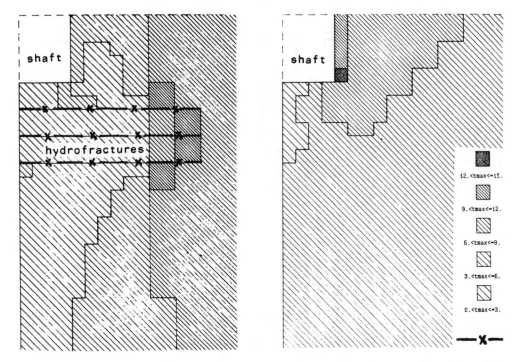

Fig.6 Max. shear stresses

The effect of partial vertical opening of artificial fractures should be helpful to this process, as well. In such a manner the above mentioned type of destressing seems to be promising as the effective measure for gradual gas releasing from outburst prone rocks.

A comparative FEM models seeking for the same destress effect made in other way were realized. Rock mass properties were modified in subjected area in several steps looking for adequate changes in the stress field there. Elastic modulus and Poisson ratio were changed in such a manner that values of Lamee's constans λ and μ were decreased on 75.50 and 25% respectively. The comparison of the models showed that approximately the same character of the stress field as according to the new concept was reached by the model with λ, μ reduction up to 25%. One dares to say that such modification of rock mass properties by using classical destress blasting is impossible to achieve, practically.

Further series of mathematical models were done for modelling of different shape of the shaft floor. There types of shapes were examined as it could be seen from Fig. 7. The types II and III were the subject of the interest and the type I was prepared for comparison as an example of normal floor shape. The interpretation showed that the type III gives better results namely in remarkable destress effect in rock "core" below the floor.

So it could be concluded that the combination of subhorizontal large scale fracturing and perimetrical presplitting anable to decrease the stress drop on newly formed boundaries by shaft sinking. These are conceptual considerations supported by some theoretical analysis. Oriented hydrofracturing – e.g. according to Chernov and Kju (1988)- seems to be the best way of large scale fracturing. Reasonable deep curfes for perimetrical presplitting could be done by using water jet technology which is developing rapidly now.

163

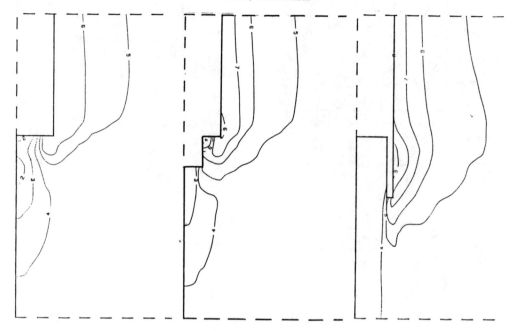

a) MAX, SHEAR STRESS

Fig.7   Influence of the floor shape

7 CONCLUSIONS

- It is necessary to cause significant volumetric destress effect
in comparatively large area below
the floor of the shaft to avoid
very intensive outbursts of rocks
and gases,
- the efficiency of destress
blasting is
limited in this matter,
- the concept based on combination
of the system of large scale
subhorizontal fracturing below
the shaft floor and perimetrical
presplitting on the floor seems
to be promising as the possible
new way of outburst prone rock
mass treatment,
- oriented hydrofracturing and
water jet technology seems to be
potencial means or the realiza-
tion of new concept,
- both the principles of the new
method and the technical means
should be the subjects of further
research.

REFERENCES

Chernov, O.I. and Kju, N.G. 1988.
Fracturing of rock massives by
fluids. J. of Phys. and Techn.
Principles of the Exploit of
Miner. Sources. 81-91 (In
Russian).
Dostál, Z. 1990. Numerical
modelling of jointed rock with
large contact zone. Proceed. of
the Int. Conf. on Mechanics of
Jointed and Faulted Rock.
595-597. A.A .Balkema, Rotterdam
Rakowski, Z. et al. 1988. The
system of prognosis of the stress
changes of field in rock massives
influenced by mining. Report.
Mining Institute of Czech. Acad.
of Sci. Ostrava. (In Czech).

Effects of Geomechanics on Mine Design, Kidybiński & Dubiński (eds) © 1992 Balkema, Rotterdam. ISBN 90 5410 040 0

Rock pressure control at longwall faces by using the powered supports

A.A.Orlov
Research Institute for Geomechanics and Mine Surveying (VNIMI), S. Petersburg, Russia

ABSTRACT: The paper contains experimental studies of rock pressure control at longwalls with using the powered supports. Consideration is given to the character of interaction of the powered support and roof rocks during working face cycle. Physical model of their interaction is shown.

INTRODUCTION

It must be said that in our country the geological conditions of coal seams occurrence show great variety in thickness, dip angle, depth, enclosing rock properties, ect. Gently sloping coal seams with a dip angle up to 35° are prevailing they account for about 77% of the total. In 80% of active longwall faces within the inclined coal strata one employs complexes with powered supports.

DISCUSSION

For the designated purpose of the powered support performance it is essential to know the regularities of its interaction with enclosing roof-and-floor rocks as well as a certain relationship of rock pressure effects to the support parameters.

On the basis of studies conducted it was assumed that in terms of the variety of roof rocks according to their loading characteristics they may be classified into two types - light-weight roofs and heavy ones. To light-weight roofs relate those roofs having great (more than 4-fold) thickness of easily caving rocks, prevailingly, argillites, siltstones, shales. Heavy roofs are characterized by existence in the vicinity of coal seam, at a distance (less than twice its thickness) of a hard

$(\sigma_{compr.} \geqslant 50$ MPa), thick (more than double thickness of a seam) layer of sandstone or limestone.

Our Institute (VNIMI) has conducted a series of specific experiments on the sequential control of powered support reaction; as a result, it was established that light-weight roof subsidence depends on the support resistance (Fig.1). This function was defined under given initial thrust of the support at the level of 100 kN/m^2 - 150 kN/m^2.

Figure 2 shows the relationships of incremental real resistance of the support for a cycle, ΔR (curve 1), final resistance, $R_f.$, (curve 2) and mean resistance, $R_m.$, (curve 3) to its initial thrust, $R_{th}.$, with its sequential changes from 70 kN/m^2 to 300 kN/m^2 and with constant nominal support resistance, $R_n.$, of 420 kN/m^2. The increment of support resistance for a cycle with the increased initial thrust drops. So, when the initial thrust is 70 kN/m^2, then the increment of support resistance accounts for 260 kN/m^2, but with the initial thrust of 300 kN/m^2 it reaches only 110 kN/m^2. At the same time, the support reaction towards the end of a cycle will increase with initial thrust rise. With the thrust of

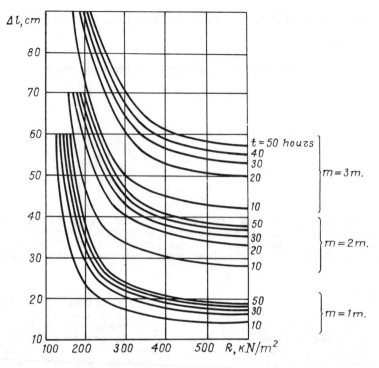

Fig.1 Light-weight roof subsidence versus the powered
support resistance with different thickness of coal
seam in time intervals.

70 kN/m^2 it reaches the value of
330 kN/m^2, with 300 kN/m^2 - 410

kN/m^2,respectively. With rise of
initial thrust the mean support
resistance will also increase during a cycle. When the initial
thrust rises from 70 to 300 kN/m^2,
then the mean support resistance

increases from 205 kN/m^2 to 355

kN/m^2. With rise of initial thrust
of the support at fixed nominal
resistance the roof subsidence
will decrease. The increment of
support resistance for a cycle
reduces as well.

In order to find out the physical substance of the regularities
established, consider the character of interaction of the powered
support with roof rocks during
a cycle from the moment of support
section installation up to its
unloading.

Figure 3 shows typical plots of
roof subsidence and support resis-

tance for a cycle, velocity of
roof subsidence during a cycle
as well as chronogram of operations
at the longwall face.

To continue, we consider the roof
behaviour and its interaction with
powered support for a cycle. Further the discussion concerns three
typical periods: a) absence of production effects in longwall face
at the site of observations (duration from 50 min. to 115 min.,Fig.
3); b) influence of coal extraction
(20 min.- 35 min., 105 min.-125 min)
c)influence of unloading and movement of support sections (38 min.-
50 min. and from 140 min.up to the
end of a cycle).

With absence of production
effects within the longwall face
at the site of observations the
roof will subside slowly, with low
almost constant rate. During coal
extraction within the area of its
influence, $l_{extr.}$,the rate of

roof subsidence will sharply rise;
the same will occur during unload-

166

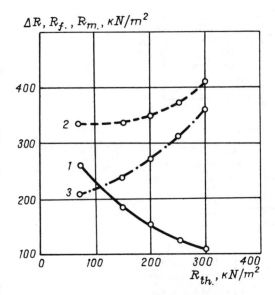

$\Delta R, R_f., R_m., \kappa N/m^2$

Fig.2 Support resistance is plotted against the initial thrust of the powered support.

ing and movement of support sections within the area of support influeence l_s.

After its movement, the support unit is installed at a new place until the resistance of initial thrust, $R_{th.}$, then it rises rather actively; and due to increased support resistance the roof subsidence rate decreases as well as the intensive resistance of the support also falls. Then, the rate of roof subsidence comes up to its steady-state value, the influence of support unloading ceases: to this time the support unit produces the resistance, $R_{set.}$; the

roof subsidence from a section setting up to this moment will be $\Delta l'_{set.}$. Then one can observe a small, gradual rise of unit's resistance up to the level of $R_{abs.}$,

caused by roof subsidence in the period of absence of production effects at the site of observations, $\Delta l'_{abs.} - \Delta l'_{set.}$. During coal extraction the support resistance substantially rises, but towards the end of coal extraction influen-

ce the support section produces the resistance $R_{extr.}$, roof subsidence for this period of a cycle accounts for: $\Delta l_{extr.} - \Delta l_{abs.}$.

Depending on the fact, to what extent the support unit's movement lags behind a mining machine the effects of next unit's unloading on the producing resistance of the observable units will take place immediately following the influence of coal extraction or later, but, in any case, at the end of a cycle one can observe the increase of support unit's resistance (if it has not reached a nominal value) up to R_f (final resistance of the support prior to its unit unloading); supplementary subsidence of roof accounts for: $\Delta l_s - \Delta l''_{set.}$.

It should be noted that the experiments conducted for the purpose of sequential changes of nominal resistance and initial thrust of the powered supports enabled to establish a number of regularities both in support resistance and roof subsidence during a cycle,

Under given initial thrust, R'_{th}, with rise of nominal support resistance (from R_{n_1} to R_{n_3}, Fig.4) its final resistance also increases from R'_{f_1} to R'_{f_3}), which reaches the nominal value ($R_{n_1} + R_{n_3}$);

in this case the roof subsidence for a cycle drops (from $\Delta l'_1$ to $\Delta l'_3$). Dependence of roof subsidence on the support resistance is of hyperbolic character (Fig.4 c). Then the moment comes when further rise of nominal resistance of the support does not lead to the rise of its final resistance. With R_{n_4}, $R'_{f_4} < R_{n_4}$, and with any $R_{n_i} > R_{n_4}$, $R'_{f_i} = R_{f_4}$, i.e. whatever high is the value of nominal resistance of the support with its initial thrust, R_{n_1}, it cannot produce its final resistance more than R'_{f_4}, nor realize the nominal resistance. Under similar conditions the support can interact with roof rocks only with-

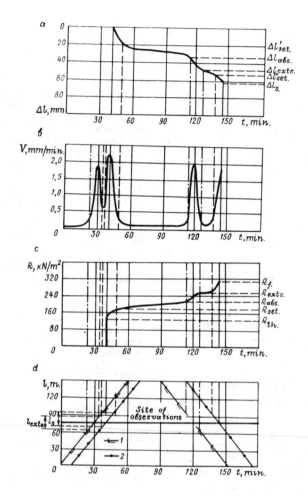

Fig.3 Graphs of roof subsidence, Δl, velocity of its subsidence, V, support resistance, R, and chronogram of operations at the longwall face during a cycle. 1 - coal cutting by mining machine; 2 - movement of support units.

in the range of resistance from R'_{f1} to R'_{f4} and roof subsidence from $\Delta l'_1$ to $\Delta l'_4$. There is no sense to have at the longwall face with similar conditions the powered support with initial thrust $R'_{th.}$ and nominal resistance $R_n > R'_{f4}$, because it will not be realized. With increased initial thrust of the support its final resistance also increases, it reaches a nominal value of greater order of magnitude than with smaller initial thrust of the support.

The initial thrust of the support

$R''_{th.} > R'_{th.}$, facilitates the progress of final resistance, $R''_f > R'_f$. and $R''_f > R_{n4}$, but $R''_f < R_{n5}$, i.e. with this initial thrust there is a certain level of nominal resistance of the support, the rise of which above this level will not already lead to the increase of final resistance. With the initial thrust, $R''_{th.}$, in the case considered the support can interact with roof only over a range of final resistance from R''_{f1} to R''_{f4} and roof sub-

168

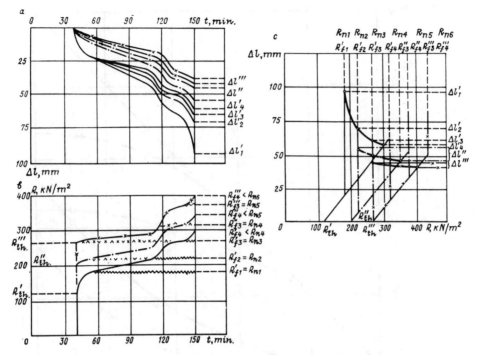

Fig.4 a) Roof subsidence; b) support resistance during a cycle, their interaction under different levels of nominal resistance, R_n., and c) initial thrust of the support, R_{th}.

sidence $\Delta l''$. Dependence of $\Delta l''$ on R''_f. with initial thrust R''_{th}. is also of hyperbolic character, as with the initial thrust R'_{th}., but now it is expressed not very clearly.

For the same range of values of final resistance, R_f. (120 kN/m²: from 180 to 300, from 230 to 350 kN/m², see Fig.4) the range of values of rock subsidence $\Delta l''$ with thrust R''_{th}. is substantially less than $\Delta l'$ with R'_{th}.

Further increase of initial thrust of the support in the case treated leads to growth of final resistance, but again up to a certain limit (see Fig.4 : with R'''_{th}. R'''_f. > R''_f. and R'''_f. > R_{n5}, but R'''_f. < R_{n6}); however, rock subsidence occurs rather slightly.

Beginning from a certain level,

the increase of initial thrust of the support leads only to the rise of its final resistance, practically without decreasing roof subsidence.

From the above it can be noted, that to apply at similar longwall face the powered support with initial support higher than the specified value is not only senseless, but harmful, because high initial thrust and high final resistance of the support will lead to rock failure at the contact with support with no minimizing rock pressure effects.

As an example, Fig.5 shows a graph of actual resistance against nominal resistance of the powered support M-88 being used at coal mine "Daryevskaya". Safety valves of the support were set up on 3 levels of nominal resistance : 450 kN/m², 240 kN/m², 180 kN/m². Further the discussion concerns the level of support resistance being produced towards the end of a cycle as well as the difference between the nominal value and real

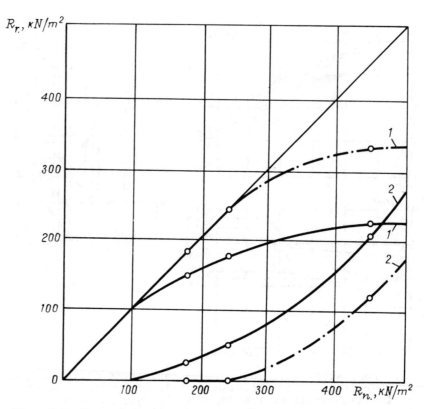

Fig. 5 Real resistance of the support versus its nominal resistance.

value available at the end of a cycle.

It can be seen from Fig. 5 that bisector of an angle characterizes an adequate employment of nominal resistance of the support, and the resistance of valve setting is equal to real resistance of the support at the end of a cycle. A plot of peak values of real resistance of the support, R_f, coincides with bisector up to the value of nominal resistance of 240 kN/m^2. With higher nominal resistance of the support its factual peak values will increase, but they do not reach already a nominal value, and the difference between them will rise with the increase of a nominal value. Mean value of the support resistance produced towards the end of a cycle becomes less than a nominal value beginning from 100 kN/m^2. With the support resistance value of

460 kN/m^2, the values of the employed and unemployed resistance become equal and further, with rise of nominal value, the employed resistance will remain nearly at the same level, but unemployed one will sharply increase. When the rise of nominal resistance of the support reaches the value more than 460 kN/m^2 with the same initial thrust, it becomes senseless since it can not be realized under these conditions.

For the reasons of the regularities established as to the support — roof interaction the investigations have been conducted concerning the roof rock movement above the productive working with the use of deep-seated reference points.

Fig.6 shows typical graphs of roof subsidence with using the deep-seated reference points within the 6th southern longwall at the coal mine "Rodinskaya". This **figure** shows also the position of reference points within a borehole as well as the chronogram of mining machine travel, unloading and movement of support units at the longwall. The graphs characterize the reference point lowering at different support resistance: a) 500 kN/m^2; b) 270 kN/m^2; c) 220 kN/m^2. It can be seen from the figure that the whole thick series from borehole log participates in movement above the working space. The value of rock movement decreases with removal from the seam at the cost of foliation. Support resistance has a substantial influence on the roof movement. When the support resistance decreases, then the lowering of all reference points begin to grow, the difference in the lowering of adjacent reference points also increases as well as roof rock foliation intensifies. The difference in values of the lowering of the last reference point positioned into sandstone, at different steps of support resistance evidences now that the support exerts an influence both on the rocks of immediate roof represented by mudstone and siltstone and on the sandstone layers of main roof.

The values of subsidence of various roof layers principally differ from each other during influence of coal extraction, unloading and movement of support sections. In the period of coal extraction (20-40 min.and 360-380 min., see Fig.6) all the reference points positioned at different distance from a seam, displace nearly at the same extent, which practically does not depend on the support resistance. In the period of unloading and movement of support units /40-60 min. and 380-400 min., Fig.6) the lowering of reference points will decrease with growth of their distance from a seam,and with the support resistance reduction it will rise. The mentioned features of roof-support interaction in various periods of a cycle can be explained by the following situation. In coal extraction one removes an abutment, the resistance of which is essentially higher of potential support resistance; the support interacts with roof rocks under conditions of speci-

fied deformation. During unloading and movement of the support that rock mass comes into a movement state which was exerted by the support influence, and rock movement will cease when the support can produce the required resistance.

In compliance with the regularities established one can represent a physical model of the powered support-roof interaction as in Figure 7. The roof is a multi-stage system; Q_i is mass of rocks in stages. Sections of the stages positioned above the working space, interact with rocks ahead of longwall face and failured rocks. They are under action of thrusting forces, R, and friction, F. Rocks in lower stages may be in the condition,when no interaction exists; first of all, they define the specified loading on the support. Each stage can move both independently with presence of foliation cavities and interacting with adjacent stages. Thus, the roof - support system being characterized by mass of rocks, Q, and support resistance, R_s., is not in closed form; in its equilibrium condition the thrusting forces, R, and friction, F, also take part, i.e.

$$R_s = \Sigma Q_i - \Sigma F_i - N,$$

where N is a responce of caving rocks. As the experiments have shown, with roof subsidence the forces R and F will rise (from Fig.4 one can see: the greater roof subsidence requires lower resistance of the support for its proper supporting. This factor defines the roof-support interaction. The whole rock mass is rested at coal seam ahead of face and caving rocks. Parallel with the movement of individual stages of roof rocks related to unloading and movement of the support units , during coal extraction the displacement of the whole rock mass takes place; this factor, first of all, will define the mode of prescribed deformation. In this way the support can produce rather different resistance depending on the value of specified deformation of roof and characteristics of the roof itself. A model of roof-support interaction shown at Fig.7 is rather sche-

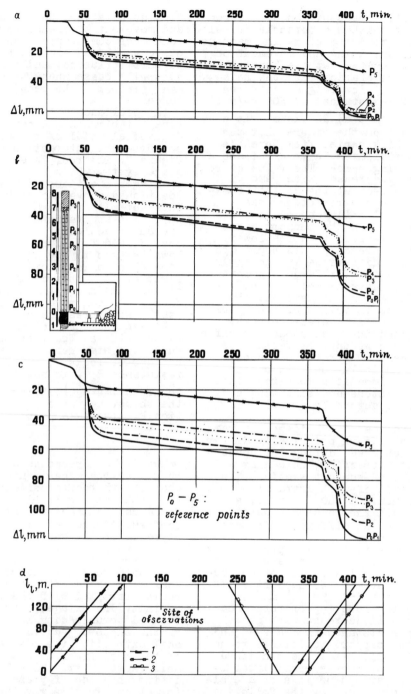

Fig.6 Graphs of roof subsidence against different support resistance with the use of reference points:
a) 500 kN/m^2; b) 270 kN/m^2; c) 220 kN/m^2.
1 - coal extraction by mining machine; 2 - movement of support units ; 3 - travel line for mining machine.

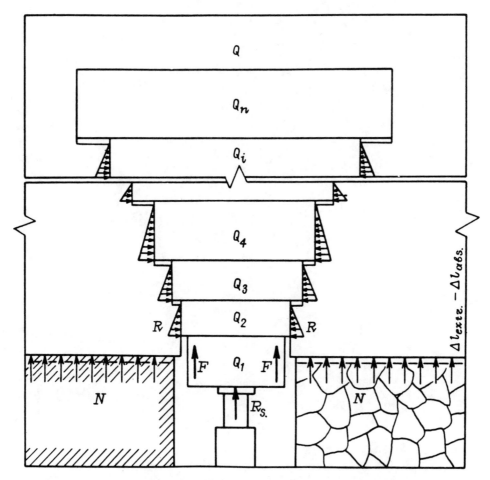

Fig. 7 A model of powered support - roof interaction.

matic. For example, the modes of prescribed deformation and its rate can be attributed to the movement of individual stages or a group of them, etc.; but, however, it gives a general representation on a character of interaction as well as explains the regularities established.

CONCLUSIONS

From the above it can be concluded that to employ the powered support with high resistance and initial thrust under all conditions seems to be rather unwise and economically unjustified.

Optimal parameters of the powered support and, first of all, its re-sistance and initial thrust, should be selected on the basis of relations of the condition and subsidence of roof to these parameters provided the best economic return.

In the USSR the following values for the resistance of powered supports have been adopted: for coal seams with light-weight roof having thickness up to 1,2 m - 300 kN/m^2, from 1,2 m to 2,5 m - 400 kN/m^2, from 2,5 m to 4,0 m - 500 kN/m^2 and more than 4,0 m - 600 kN/m^2. For coal seams with heavy roofs the values of support resistance are adopted twice as much. Initial thrust of the supports accounts for from 60% to 80% of their nominal resistance.

Effects of Geomechanics on Mine Design, Kidybiński & Dubiński (eds) © 1992 Balkema, Rotterdam. ISBN 90 5410 040 0

Author index